# 50
# Topics in Heterocyclic Chemistry

## Aims and Scope

The series Topics in Heterocyclic Chemistry presents critical reviews on present and future trends in the research of heterocyclic compounds. Overall the scope is to cover topics dealing with all areas within heterocyclic chemistry, both experimental and theoretical, of interest to the general heterocyclic chemistry community.

The series consists of topic related volumes edited by renowned editors with contributions of experts in the field.

More information about this series at http://www.springer.com/series/7081

Philipp Selig
Editor

# Guanidines as Reagents and Catalysts I

With contributions by

D.H. Grayson · T. Ishikawa · M. Kato · Y. Kobayashi · D. Leow · K. Nagasawa · C. Nájera · M. Odagi · I. Rozas · J.W. Shaw · Y. Takemoto · C.-H. Tan · M. Yus

Springer

*Editor*
Philipp Selig
Patheon Inc.
Linz, Austria

ISSN 1861-9282 ISSN 1861-9290 (electronic)
Topics in Heterocyclic Chemistry
ISBN 978-3-319-52723-9 ISBN 978-3-319-52725-3 (eBook)
DOI 10.1007/978-3-319-52725-3

Library of Congress Control Number: 2017936046



Printed on acid-free paper

This Springer imprint is published by Springer Nature
The registered company is Springer International Publishing AG
The registered company address is: Gewerbestrasse 11, 6330 Cham, Switzerland

# Preface

Guanidines, the all-aza analogues of carbonic acids, represent a fascinating group of molecules with unique chemical and physical properties. Just as the well-known amidines, guanidines are exceedingly strong Brønsted bases and are therefore even referred to as "superbases." Moreover, guanidines can exhibit strong Lewis-basic properties and thus serve as electron-pair donors and ligands. After protonation, the highly stabilized guanidinium cation is often used as a powerful, bidentate H-bond donor, capable of tight binding and activation of a variety of H-bond acceptors such as carbonyl groups. Finally, guanidinium cations can also be regarded as Lewis-acidic species which can act as π-Lewis acids.

Guanidines and their corresponding protonated species are thus capable of exhibiting all four basic chemical functionalities: free bases are Lewis and Brønsted basic, while cations are Lewis and Brønsted acidic, all connected by a simple proton transfer.

Besides this obvious potential for synthetic applications, guanidines are also a challenging target for synthetic endeavors, mainly due to their highly basic character. In the first volume of *Guanidines as Reagents and Catalysts*, we thus wanted to open with an overview of *Prof. Rozas*, which introduces the reader to principal techniques for guanidine synthesis and offers a first glimpse on the potential of guanidines in biological applications.

A main topic of Vol. I concerns the use of guanidines as synthetic reagents or, more specifically, as organocatalysts. We are introduced into this topic with a chapter by *Prof. Ishikawa,* a pioneer of guanidine organocatalysis and also the inventor of one of the very rare examples of a commercially available guanidine catalyst, "Ciba-G." Ciba-G is also already highlighting the importance of multi-functional activations in guanidine organocatalysis, a most important concept, which is further illustrated by the works of *Prof. Takemoto* in the following chapter.

Turning the focus from catalyst structures to synthetic applications, *Prof. Najera* will elaborate on a pivotal guanidine-catalyzed reaction, i.e., the Michael addition, which makes formidable use of both the Brønsted basic and the H-bond donating properties of the guanidine and guanidinium cation. In the following chapter, structures of guanidine organocatalysts are taken to the next level by *Prof. Tan* and his introduction of bicyclic guanidine organocatalysts. These synthetically useful, as well as aesthetically pleasing structures show us that highly efficient catalysis may not be strictly contingent upon multifunctional activation, and steric effects around an isolated guanidine moiety can be sufficient to achieve excellent results. While Prof. Tan's work focuses on sterically rigid, mono-functional catalysts, a quite antipodal approach, using highly flexible guanidines with multiple functional groups attached, is shown to succeed just as well in the final chapter of Vol. I by *Prof. Nagasawa*.

While the majority of Vol. I deals with guanidines as reagents and catalysts in the field of organic synthesis, the potential uses of guanidines certainly go far beyond that. In the second of these two volumes a focus is placed on the specialized applications of guanidines. In the first chapter *Prof. Concellón* and *Prof. del Amo* show us their works on structurally simple guanidinium salts to effectively modify reactions catalyzed by the classic organocatalyst L-proline, demonstrating the design of elaborate new catalyst structures is not necessarily mandatory to benefit from guanidine catalysis. Guanidine organocatalysis is also involved in an industrially useful field, namely the nucleophilic activation of $CO_2$ as a sustainable C1-building block. *Prof. Pérez González* presents this "green" use of guanidine catalysis in the following chapter.

Further highlighting the potential uses of guanidines outside traditional organic synthesis *Prof. Oppel* presents guanidines as ligands for super-molecular metal-based frameworks, and the synthetic potential of such guanidinium-metal complexes is explored by *Prof. Herres-Pawlis*, exemplified in their use as highly active polymerization catalysts. Finally, at the end of this volume, *Prof. Himmel* takes us far beyond our focus on synthetic organic chemistry with his chapter on the unique electronic properties of anionic guanidinates and their complexes.

In summary, it was our goal to show that guanidines, guanidinium salts, and guanidinates offer a very diverse range of reactivity and thus great potential for a wide variety of uses. While still being regarded as a rather exotic class of molecules in the field of organocatalysis, especially in comparison to the prominent field of proline-induced imine/enamine activation or H-bond catalysis enabled by thioureas, the potential of guanidines as reagents and catalysts as well as the door to novel applications is certainly wide open. Currently, in the mid-2010s, guanidine chemistry is a highly dynamic and rapidly developing field of research, and we can expect exciting new developments in the future. Guanidines as reagents and catalysts are here to stay and will continue to show up as versatile and valuable tools both in and beyond organic chemistry.

Linz, Austria
November 2016

Philipp Selig

# Contents

Top Heterocycl Chem (2017) 50: 1–52
DOI: 10.1007/7081_2015_174

Published online: 20 November 2015

# Synthesis of Guanidines and Some of Their Biological Applications

**Julian W. Shaw, David H. Grayson, and Isabel Rozas**

**Abstract** Guanidine is one of the most versatile functional groups in chemistry; compounds containing this system have found application in a diversity of biological activities, and in this chapter, the advances in the field of the synthesis of guanidines are presented. First, the preparation of acyclic guanidines involving the reaction of an amine with an activated guanidine precursor followed by the deprotection to yield the corresponding free guanidine is discussed. Thiourea derivatives as guanidylating agents have been widely used as guanidine precursors using coupling reagents or metal-catalysed guanidylation. Alternatively, *S*-methylisothiourea has shown to be a very efficient guanidylating agent, and *N*,*N*′,*N*″-trisubstituted guanidines have also been used to install the guanidine functionality. Despite the similarity between urea and thiourea, the former has received much less attention; however, its application in guanidine synthesis has also been proved. Examples of the preparation of guanidines using cyanamides that react with derivatised amines as well as the use of copper-catalysed cross-coupling chemistry are also presented. Moreover, cyclic guanidines such as 2-aminoimidazolines (five-membered rings), 2-amino-1,4,5,6-tetrahydropyrimidines (six-membered rings) and 2-amino-4,5,6,7-tetrahydro-1H-1,3-diazepines (seven-membered rings) are present in many natural products and compounds of medicinal interest. Accordingly, an overview of the methods found in the literature for the preparation of these cyclic guanidines is presented. Finally, some biological applications of guanidines as DNA minor groove binders, kinase inhibitors and $\alpha_2$-noradrenaline receptors antagonists are discussed.

J.W. Shaw, D.H. Grayson, and I. Rozas (✉)
School of Chemistry, Trinity Biomedical Sciences Institute, Trinity College Dublin, 152-160 Pearse Street, Dublin 2, Ireland
e-mail: rozasi@tcd.ie

**Keywords** Guanidine · Synthesis

## Contents



# Abbreviations

| | |
|---|---|
| $\alpha_2$-AR | $\alpha_2$-Adrenoceptor |
| Boc | *tert*-Butyloxycarbonyl |
| Cbz | Carboxybenzyl |
| CSA | Camphor sulfonic acid |
| Dba | Dibenzylideneacetone |
| DCC | Dicyclohexylcarbodiimide |
| DDQ | 2,3-Dichloro-5,6-dicyano-1,4-benzoquinone |
| DEAD | Diethylazodicarboxylate |
| DIPEA | Diisopropylethylamine |
| DMAP | 4-Dimethylaminopyridine |
| DME | Dimethoxyethane |
| DMF | Dimethylformamide |
| DMP | Dess–Martin periodinane |
| DMSO | Dimethyl sulfoxide |
| DNA | Deoxyribonucleic acid |
| DtBPF | 1,1′-Bis(di-*tert*-butylphosphino)ferrocene |
| EDCI | 1-Ethyl-3-(3-dimethylaminopropyl)carbodiimide |
| Fmoc | Fluorenylmethoxycarbonyl |

| | |
|---|---|
| Gua | Guanidine |
| HB | Hydrogen bond |
| LDA | Lithium diisopropylamine |
| LiHMDS | Lithium hexamethyldisilazide |
| MBH | Morita–Baylis–Hillman |
| MGB | Minor groove binder |
| mIBG | *meta*-Iodobenzylguanidine |
| Mtr | 2,3,6-Trimethyl-4-methoxybenzenesulfonyl |
| NBS | *N*-Bromosuccinimide |
| NIS | *N*-Iodosuccinimide |
| NMM | *N*-Methylmorpholine |
| PET | Positron emission tomography |
| Ph | Phenyl |
| PhNO | Nitrosobenzene |
| Piv | Pivaloyl |
| PMB | *p*-Methoxybenzyl |
| Pmc | 2,2,5,7,8-Pentamethylchroman-6-sulfonyl |
| Py | Pyridine |
| PyHBr3 | Pyridinium tribromide |
| RNA | Ribonucleic acid |
| RSM | Recovered starting material |
| TBAB | *tetra*-Butylammonium bromide |
| TBD | 1,5,7-Triazabicyclo[4.4.0]dec-5-ene |
| Tces | 2,2,2-Trichloroethoxysulfonyl |
| TCT | 2,4,6-Trichloro-1,3,5-triazine |
| TFA | Trifluoroacetic acid |
| TFAA | Trifluoroacetic anhydride |
| THF | Tetrahydrofuran |
| TIPS | Triisopropylsilyl |
| TMEDA | Tetramethylenediamine |
| TMG | 1,1,3,3-Tetramethylguanidine |
| TON | Turnover number |
| Tosyl | 4-Toluenesulfonyl |
| Trifyl | Trifluoromethylsulfonyl |
| Troc | 2,2,2-Trichloroethoxycarbonyl |
| TsCl | 4-Toluenesulfonyl chloride |
| Xantphos | 4,5-Bis(diphenylphosphino)-9,9-dimethylxanthene |

## 1 Introduction

The ability of guanidines to form hydrogen bonds (HBs), their planarity and their high basicity are some of their predominant characteristics that make this functional group a very versatile one for compounds with biological activity. The high basicity

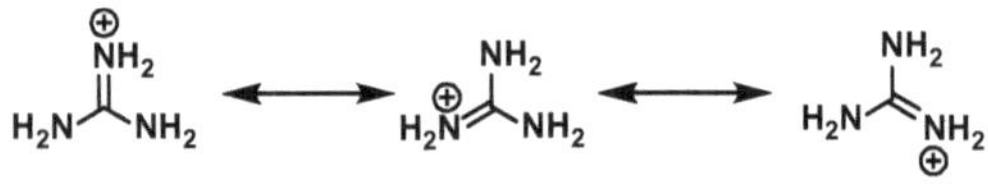


**Fig. 1** Guanidinium resonance forms

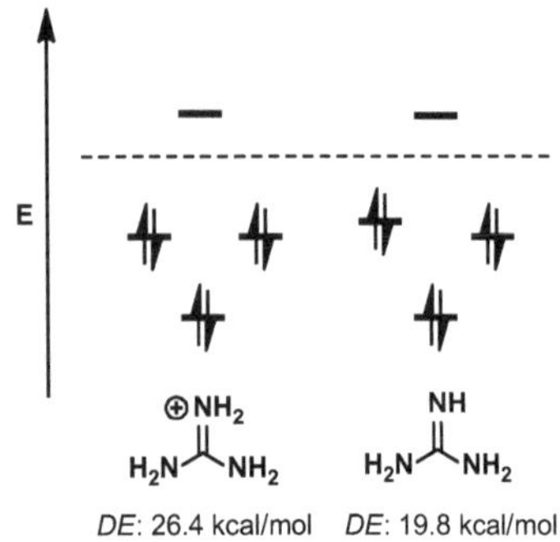


**Fig. 2** Electronic configuration (Hückel molecular orbitals) of guanidinium cation and guanidine. *DE* delocalisation energy [10]

of guanidine (the $pK_{aH}$ of the guanidinium cation in water is 13.6 [1]) indicates that at physiological pH, guanidine will be protonated forming the guanidinium cation. Due to the conjugation between the lone pairs of the nitrogen atoms and the imine double bond, protonated guanidines have a number of resonance forms that delocalise the positive charge over the entire functional group leading to its high basicity (Fig. 1). This feature of the guanidinium cation causes the nitrogenous backbone to be planar, and this can often determine the conformation of substituted guanidinium species as well as the interaction with aromatic systems in biological environments such as amino acids and nucleic acid bases [2, 3].

In the guanidinium moiety, six protons are available for hydrogen bonding interactions. This allows the critical base-pairing interactions that are possible for guanine with cytosine in DNA [3] and also facilitates numerous arginine interactions throughout the mammalian body [4, 5]. Being protonated at physiological pH, the guanidine moiety in arginine will be cationic and planar. Besides being a HB donor, arginine is able to interact with other molecules through weaker interactions such as cation-π interactions [6–9]. The guanidinium cation has 6π electrons and all of them are in the bonding orbitals (Fig. 2) [10].

This gives the guanidinium ion a type of electron delocalisation that in some cases is referred to as “Y aromaticity”, allowing interactions with planar electron-rich moieties such as phenyl rings [11, 12]. Considering that these properties of guanidine make it an important functional group in both the biological [13] and chemical [14] sciences, the efficient synthesis of guanidine-containing molecules is highly relevant, and, accordingly, in this chapter, methods for the introduction of the guanidine function will be discussed. This chapter aims to highlight the vast possibilities for synthesising guanidines and act as a guide to which method of guanidylation may be the most effective in a given situation.

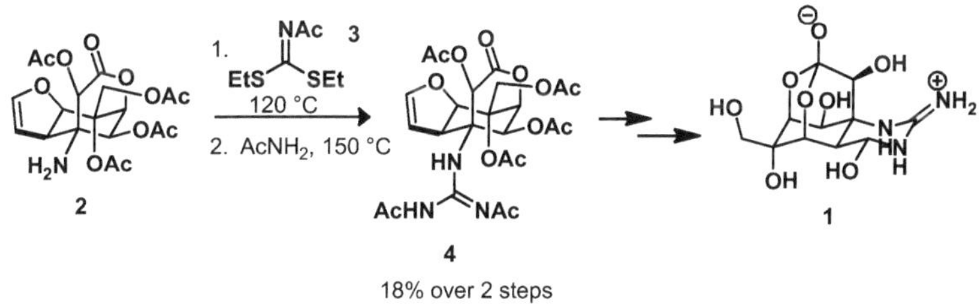


Scheme 1

Scheme 2

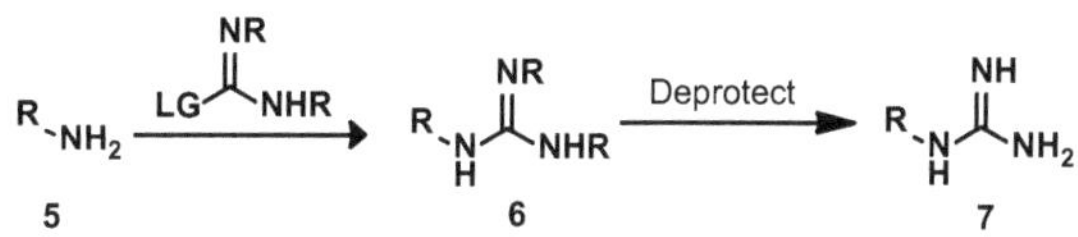


## 1.1 *Synthetic Methods for the Construction of the Guanidine Functionality*

The preparation of guanidines has come a long way from the landmark synthesis of tetrodotoxin **1** by Kishi in 1972 [15] in which the guanidine functionality was introduced by exposing amine **2** to dithiocarbonimidate **3** at elevated temperature and, then, exposing the intermediate product to acetamide ($AcNH_2$) to form the desired protected guanidine **4** in a yield of 18% over two steps (Scheme 1).

# 2 Synthesis of Acyclic Guanidines

Advances in the field have led to a myriad of methods for the installation of the guanidine functional group [16]. In the past, the synthesis of guanidines generally involved the reaction of an amine such as **5** with an activated guanidine precursor followed by the deprotection of this moiety **6** to yield the corresponding free guanidine **7** (Scheme 2).

The choice of guanidine precursor usually has depended upon the reactivity of the amine involved; in the case of anilines, Lewis acid-promoted reactions have been the most successful, whilst for aliphatic amines, there is a much wider range of guanidine precursors available due to their greater nucleophilicity. Herein, we will discuss the classical methods for the preparation of guanidine derivatives followed by more recent developments in the area.

Scheme 3

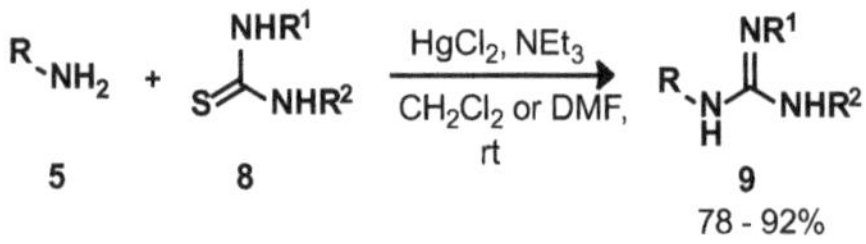


## 2.1 Thiourea Derivatives as Guanidylating Agents

Classical methods for guanidine synthesis have tended to make use of either substituted thiourea or urea moieties being reacted with an amine promoted by a variety of reagents. The earliest examples of thiourea being utilised as a guanidylating agent made use of the thiophilic Lewis acid $HgCl_2$ as a stoichiometric promoter [17]. Mercury(II) chloride is believed to act as a desulphurising agent forming an electrophilic carbodiimide which in turn is rapidly attacked when in the presence of a nucleophilic amine **5** (Scheme 3).

This methodology is extremely robust and synthetically useful. For example, the reaction of a variety of aromatic and aliphatic amines with *N,N*-bis-*tert*-butoxycarbonylthiourea **8** in the presence of $NEt_3$ and $HgCl_2$ in $CH_2Cl_2$ affords the corresponding protected guanidines **9** in good to excellent yields (Scheme 3, 78–92%). Acidic deprotection then yields the corresponding guanidinium salts. The use of HgO as a replacement for $HgCl_2$ has been shown to be a viable alternative [18].

Some of the reported applications of this useful synthetic methodology in the preparation of potentially active derivatives include the synthesis of 3-guanidinium-4′-arylguanidinium diaromatic derivatives **10** as kinase inhibitors (Scheme 4a) [19] and the preparation of 2-pyridinoguanidines **11** as $\alpha_2$-adrenoceptor antagonists (Scheme 4b) [20].

Another interesting example of the synthetic utility of mercury-promoted guanidylation is the reaction between thiourea derivatives **12** (Scheme 5) and *N*-iminopyridinium ylide, generated from compound **13**, to form the unusual *N*-pyridinium benzoylguanidines **14** [21]. Bismuth salts have also been shown to be effective in promoting this guanidylation reaction.

A useful method for the synthesis of *N,N′*-disubstituted aryl guanidines starts with Boc protection of thiourea **15** (Scheme 6) followed by exposure to trifluoroacetic anhydride (TFAA) which forms a highly reactive difunctionalised nitrogen atom [22]. This substituent can in turn be displaced by addition of a nucleophile to form derivatised thiourea **8**. Exposure of **8** to $HgCl_2$-promoted guanidylating conditions results in the formation of *N,N′*-disubstituted *mono*-Boc-protected guanidines **16** (Scheme 6).

However, the use of mercury-containing reagents in the above methodologies is an obvious drawback when aiming at the synthesis of potential therapeutic agents, and other methods have been devised to deal with this problem. Thus, in 2006, Cunha et al. described the use of Bi(III) salts to promote the desulphurisation of protected thiourea **8** which was subsequently attacked by a nucleophilic amine species **5** constructing the appropriate protected guanidine **9** (Scheme 7) with expulsion of a Bi–S species [23].

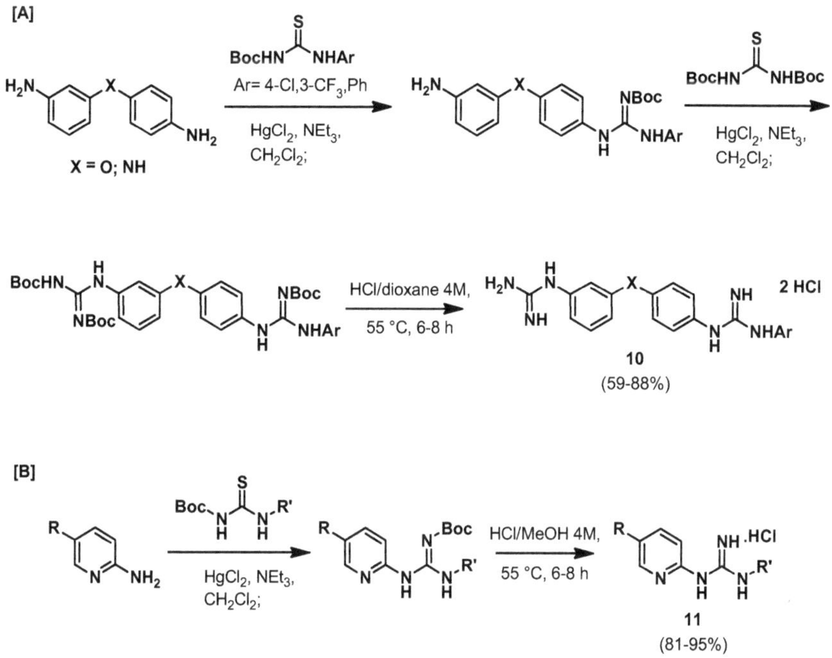


**Scheme 4**

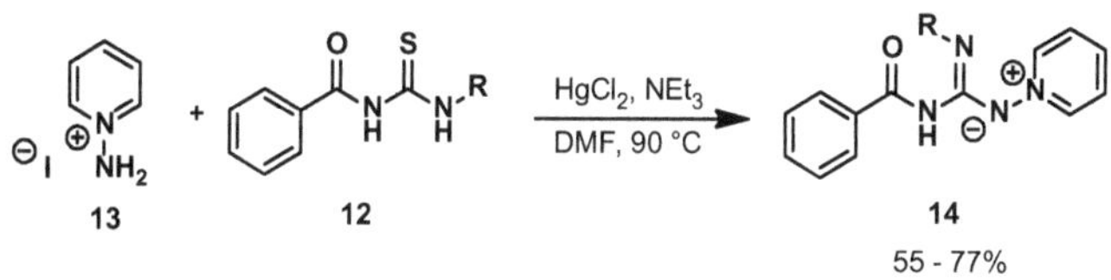


**Scheme 5**

**Scheme 6**

**Scheme 7**

**Scheme 8**

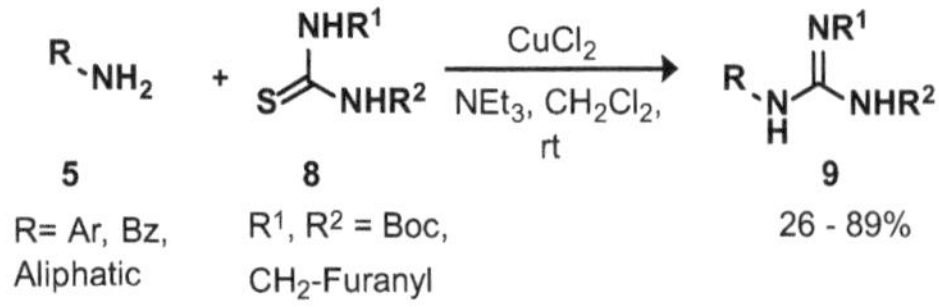


**Scheme 9**

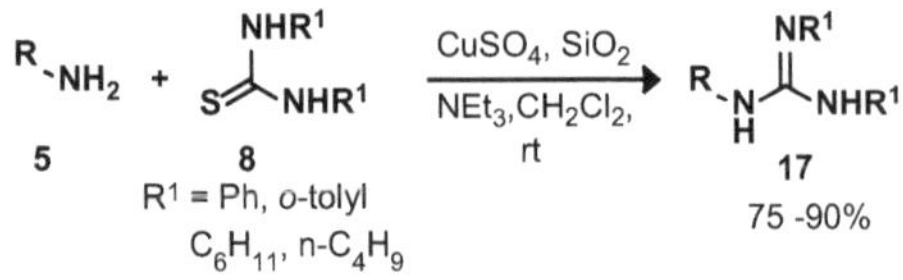


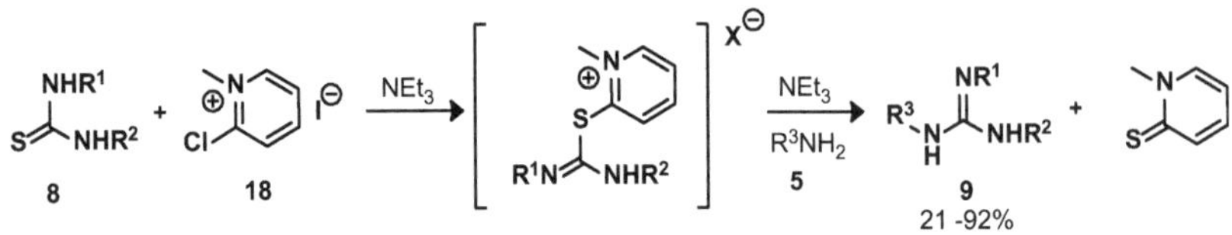


**Scheme 10**

These authors [23] found that $BiI_3$ could be used in catalytic quantities (5 mol%) when $NaBiO_3$ was used as an oxidant. This method proved to deliver yields comparable with the $HgCl_2$-promoted guanidylation reaction when either aliphatic amines or activated anilines were used. However, the substrate scope did not include deactivated amines.

The use of Cu(II) chloride salts has also been reported to effect the conversion of protected thiourea **8** into the corresponding guanidines **9**. Initial reports from Kim et al. [17] described $CuCl_2$ as being an ineffective promoter of guanidylation; however, more recent findings [24] suggest that $CuCl_2$ can be used as a Lewis acid with guanidylation yields similar to those obtained in the benchmark $HgCl_2$ reaction (Scheme 8). As expected, non-nucleophilic amines required heating and gave poor to moderate yields.

Copper(II) sulphate ($CuSO_4$) has also proven to be an active promoter in guanidylation reactions [25]. When used in conjunction with $SiO_2$ and $NEt_3$ in THF, a number of aliphatic amines could be guanidylated with substituted thiourea in adequate yields (Scheme 9). No aromatic amines were evaluated though, indicating a lack of utility for the method when applied to non-nucleophilic amines.

An attractive alternative to the utilisation of metal-promoted guanidylation is the use of Mukaiyama's reagent **18** to assist in the desulphurisation of the thiourea. Mukaiyama's reagent works best when nucleophilic amines are used (Scheme 10) [26].

It has been shown that the choice of solvent can play a critical role in the reaction of the protected thiourea **8** with varying amines when using Mukaiyama's reagent **18**. The reaction of non-nucleophilic amines with thiourea **8** was shown to be much

**Scheme 11**

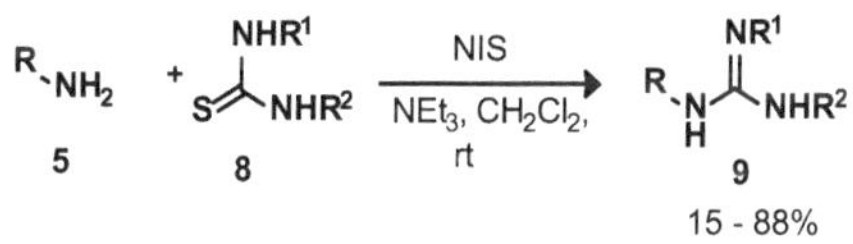


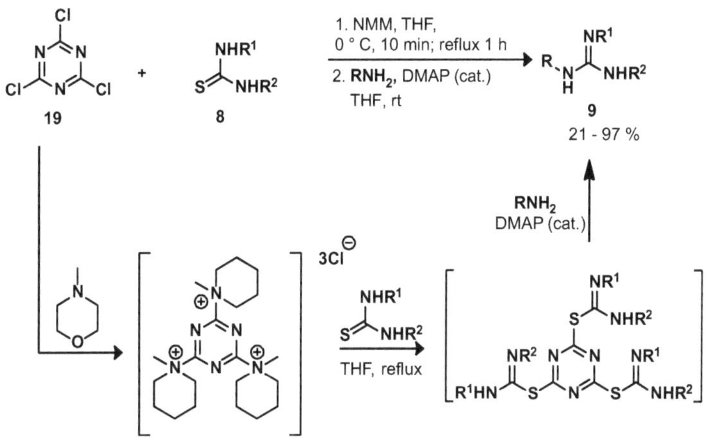


**Scheme 12**

more effective in dichloromethane than in dimethylformamide (the typical choice of solvent for this reaction), with yield increases of around 50% being achieved.

*N*-Iodosuccinimide (NIS) has also proven to be a viable choice for the guanidylation of an amine **5** with protected thiourea **8** (Scheme 11). In 2009, Smietana and co-workers reasoned that thiophilic NIS, acting as a soft Lewis acid, should coordinate to the sulphur of the protected thiourea and, in the presence of an amine and a base, form the desired guanidine **9** [27].

An array of amines was investigated, with primary aliphatic amines giving excellent yields and hindered secondary amines giving moderate yields. When bis-Boc-*S*-methylisothiourea replaced bis-Boc-thiourea as the guanidylating agent, higher yields were obtained for secondary amines.

An interesting metal-free guanidylation procedure makes use of 2,4,6-trichloro-1,3,5-triazine (TCT or cyanuric chloride **19**) as a promoter for desulphurisation of protected thiourea **8** [28]. Compound **19**, when used in conjunction with *N*-methylmorpholine (NMM) and catalytic DMAP in THF, activates bis-Boc-thiourea **8** leading to efficient guanidine formation. TCT can be used at 0.33 mol/equiv. due to the three available sites for activation (Scheme 12).

Upon exposure to sodium molybdate dihydrate ($Na_2MoO_4{\cdot}2H_2O$) and hydrogen peroxide ($H_2O_2$, 30% in water), thioureas **8** are known to be converted into sulfonic acids **21** (Scheme 13) [29].

Maryanoff et al. discovered that these sulfonic acid substrates **21** are viable options as guanidylating agents [29]. Upon exposure to nucleophilic amines, the oxidised sulphur can be displaced to yield the desired guanidine **9** (Scheme 13). The reaction of several monosubstituted thioureas under these oxidising conditions was

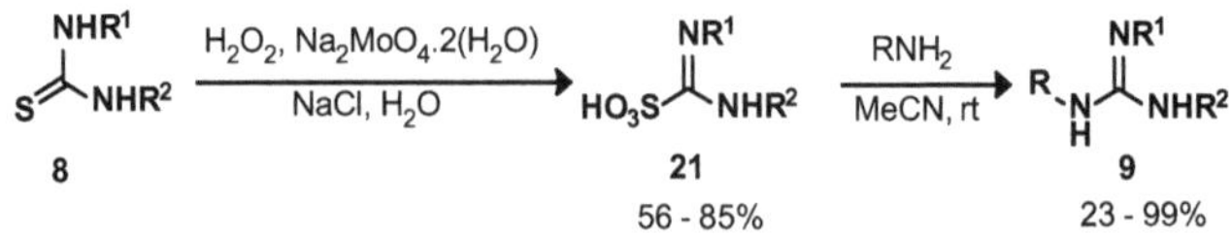


Scheme 13

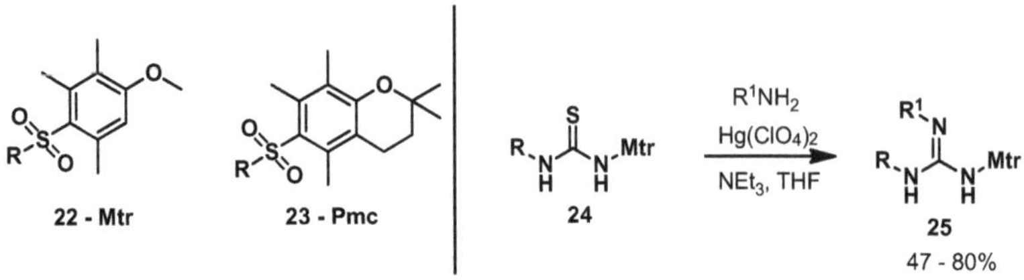


Scheme 14

reported to give good yields, and these sulfonic acids **22** were shown to be thermally stable at room temperature. A range of amines was investigated for their efficacy in displacing the oxidised sulphur, and good to excellent yields of guanidine derivatives were reported. Elevated temperatures were required for hindered or non-nucleophilic amines.

In the search for protecting groups orthogonal to those typically used in peptide synthesis, and in particular for peptides containing arginine residues, two new functionalities have been discovered. The 2,3,6-trimethyl-4-methoxybenzene-sulfonyl (**22**, Mtr) and 2,2,5,7,8-pentamethylchroman-6-sulfonyl (**23**, Pmc) groups have been designed to withstand cleavage conditions that would typically remove either Boc, Cbz, or Fmoc protecting groups (Scheme 14) [30].

Mtr- or Pmc-protected thiourea (e.g. **24**) can be reacted with a range of amines under typical guanidylation conditions [$Hg(ClO_4)_2$ proved to be optimal] to form both Mtr- or Pmc-protected guanidines (e.g. **25**), respectively (Scheme 14). These groups, as previously alluded to, are particularly useful in peptide synthesis incorporating guanidine-containing arginine residues. Deprotection occurs under strongly acidic conditions [30].

## 2.2 *Coupling Reagents*

In a number of cases, the coupling reagent EDCI [1-ethyl-3-(3-dimethylamino-propyl)carbodiimide] has been shown to be capable of promoting guanidylation between thiourea derivatives and nucleophilic amines. In an example published in 2002, EDCI was used to effectively facilitate the desulphurisation of thiourea **26** and subsequent guanidine **27** formation (Scheme 15) [31]. Compound **26** had been

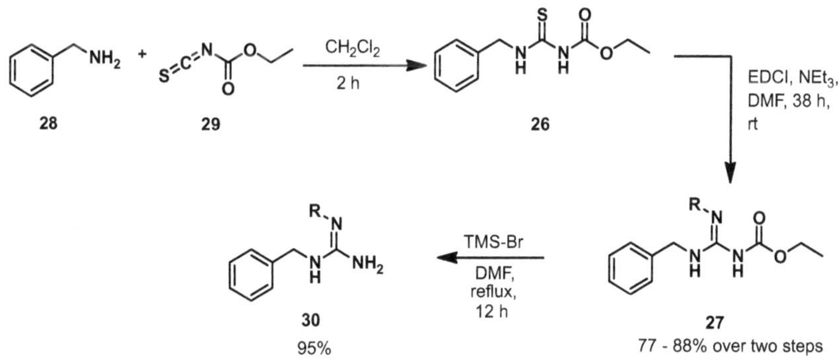


Scheme 15

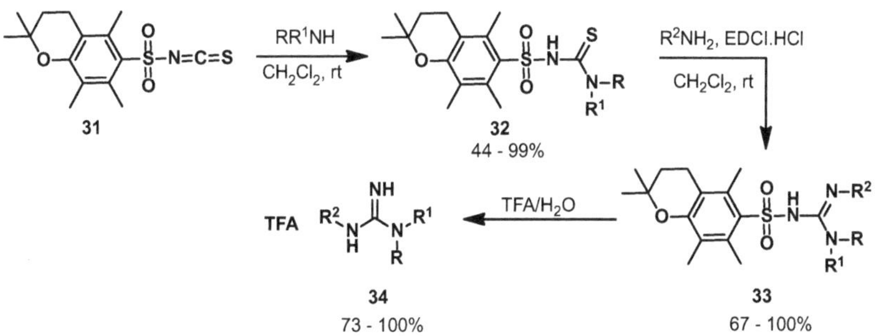


Scheme 16

synthesised by reaction of benzylamine **28** and ethyl thiocyanatoformate **29** in $CH_2Cl_2$ affording the product after 2 h in a 98% yield.

Most of the amines employed were aliphatic in nature with only one example of a non-nucleophilic amine, aniline, being used. To remove the ethyl carbamate group, typical methodologies failed [31]. Interestingly, trimethylsilyl bromide under reflux conditions in DMF turned out to be the condition of choice in this deprotection. However, only a single yield of 95% for **30** was reported for this deprotection step without a list of substrates being provided.

Another example found in the literature from 2007 also effectively uses EDCI to form substituted guanidines (Scheme 16) [32].

In the example shown in Scheme 16, *N*-Pmc-isothiocyanate **31** was reacted with a variety of both primary and secondary amines to form substituted thioureas **32** in good to excellent yields. Subsequently, other amines were reacted with the substrate **32** separately in the presence of the coupling reagent EDCI to form compounds **33**. It is important to note the viability of forming derivatised guanidines in

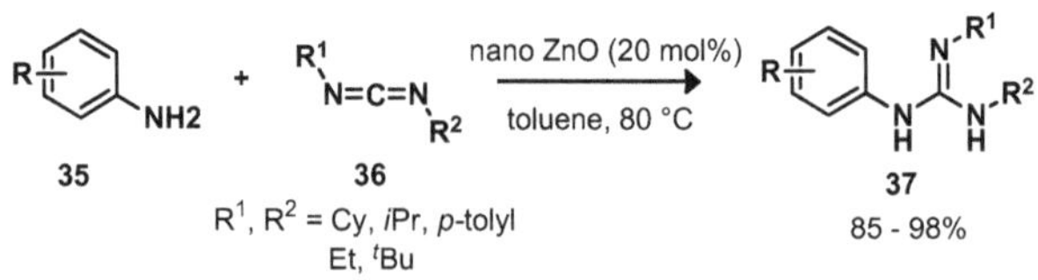


**Scheme 17**

this manner. The relatively electrophilic isothiocyanate can react with a plethora of amines of varying nucleophilicites. However, the EDCI-promoted step requires nucleophilic primary amines to produce optimal results. By choosing the appropriate amines for each step, the authors were able to synthesise a number of substituted guanidines which could then undergo deprotection of the Pmc group in acidic conditions to yield the guanidinium trifluoroacetate salts **34** [32].

## 2.3 *Metal-Catalysed Carbodiimide Guanidylation*

The reaction between an amine and a carbodiimide is one of the most atom-economical methods of synthesising guanidines and has been a fruitful area of research in recent years. Numerous examples of metal-catalysed reactions between amines and carbodiimides have been published [33]. In 2012, it was reported that aniline **35** could be reacted with dicyclohexylcarbodiimide (DCC) and its derivatives **36** in the presence of a heterogeneous metal species to form the corresponding guanidine **37**. Initial investigations (Scheme 17) showed that nanocrystalline zinc oxide (nano ZnO) catalysed this reaction effectively, and after a quick solvent screen, it was found that nonpolar, non-coordinating solvents such as toluene gave optimal results (yields up to 98%) [34].

Interestingly, the use of commercial ZnO as opposed to nano ZnO gave drastically inferior yields (~5%). The nano ZnO could be recycled up to six times, giving similar yields on each occasion without any reduction in activity. A wide variety of anilines and a number of carbodiimides could be reacted under similar conditions generating the required trisubstituted guanidines. The drawback of this chemistry, however, is that only a select few carbodiimides are stable and/or commercially available, allowing only very particular substrates to be synthesised in this way. Hindered amines required higher temperatures for reaction to occur. The ZnO is thought to activate the carbodiimide facilitating nucleophilic attack by the amine.

In another report in 2012, $Fe(OAc)_2$ (2 mol%) was shown to catalyse the reaction between anilines **35** and carbodiimides **36** just as effectively as nano ZnO and under similar conditions [35]. As in the previous report, a wide range of aromatic and aliphatic amines were guanidylated in this manner. Similarly, the drawback is the lack of available carbodiimides to obtain a truly diverse library of guanidines.

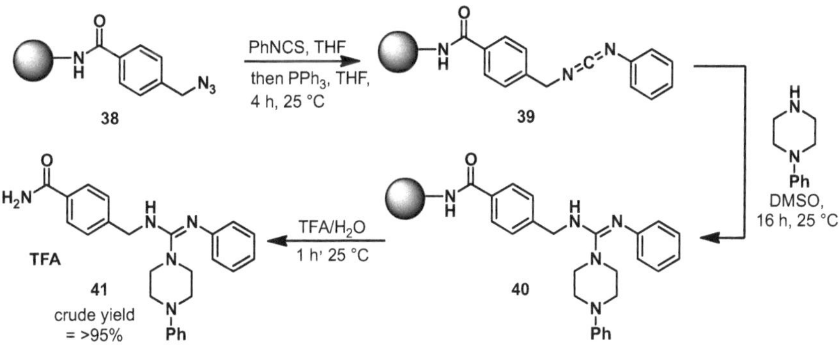


Scheme 18

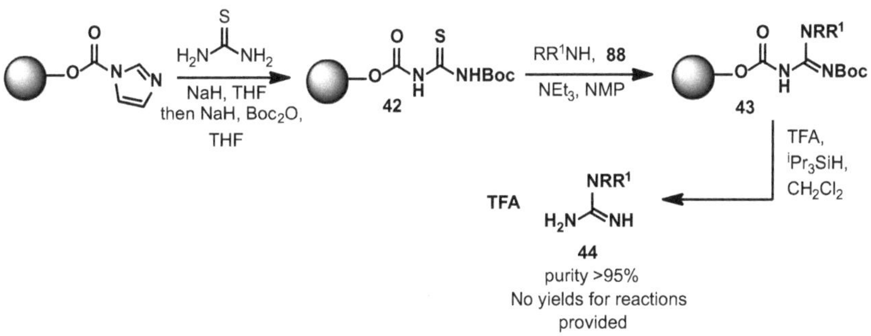


Scheme 19

## 2.4 Polymer-Supported Guanidylation

A number of methods for the solid-phase syntheses of guanidines have been reported. The use of polymer-supported reagents has a number of advantages, namely, facilitating the removal of excess reagent, unreacted starting materials and unwanted by-products from the reaction mixture [36].

The utility of solid-supported reagents in the preparation of guanidines has been demonstrated for the synthesis of *N,N′,N″*-trisubstituted guanidines in a report by Drewry et al. in 1997 (Scheme 18) [37]. The synthesis began with the preparation of solid-phase-supported reagent **38**. This azide was then exposed to $PPh_3$ to form an iminophosporane which, in the presence of PhNCS, can undergo an aza-Wittig coupling forming the prerequisite carbodiimides **39**, which are known to be stable in solid-phase chemistry. Compound **39** can now be reacted with a variety of amines such as *N*-phenylpiperazine to form polymer-bound guanidine **40**. Under acidic conditions, the desired guanidine can be cleaved from the resin to give the target compounds **41** in good yield.

The synthesis of *mono*-substituted aryl guanidines can also be achieved using solid-support chemistry (Scheme 19). The solid-supported Boc-protected thiourea

Scheme 20

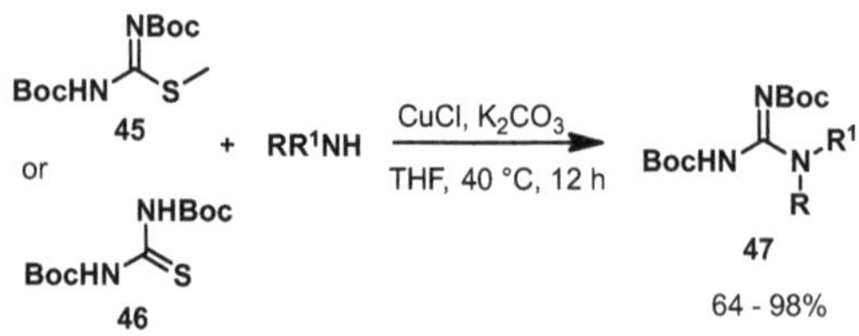


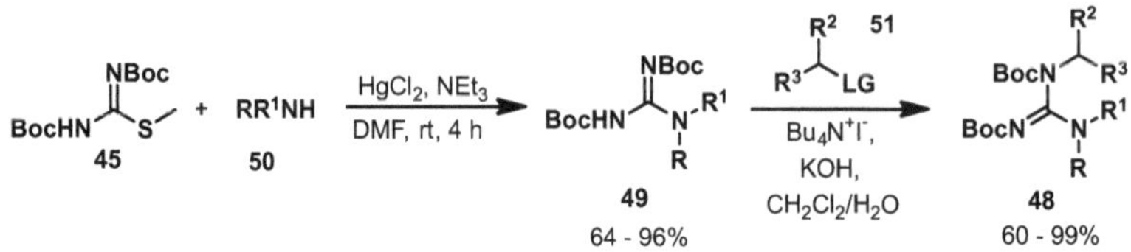


Scheme 21

**42** can be thought of as a polymer-bound guanidylating agent [38]. In the presence of a promoter such as Mukaiyama's reagent, a range of amines can be reacted with the thiourea derivative to form polymer-supported Boc-protected guanidines **43**. Upon exposure to TFA, the guanidines then are cleaved from the resin yielding the desired *mono*-substituted guanidines as their trifluoroacetic acid salts **44**.

## *2.5 S-Methylisothiourea as Guanidylating Agent*

The successful use of *S*-methyl-*N*,*N*′-bis-Boc-isothiourea **45** as a guanidylating agent has been reported in a number of literature examples. Its ability to replace *N*,*N*′-bis-Boc-thiourea **46** has proven to be advantageous in a number of circumstances. In a discovery by Rozas and co-workers [24], it was found that copper (II) salts could promote guanidylation of *N*,*N*′-bis-Boc-thiourea with a variety of amines, but either low or no yields were obtained for the analogous reaction with *S*-methyl-*N*,*N*′-bis-Boc-isothiourea. However, in a publication by Terada and co-workers [39], it was demonstrated that copper(I) salts could effectively promote guanidylation of both thiourea precursors (Scheme 20). In this work, a number of amines were exposed to both sets of conditions, and the protected guanidine products **47** were obtained in good to excellent yields (Scheme 20). This, interestingly, shows the specificity that promoters can sometimes have in guanidylation reactions and gives an insight into why there is such a diverse literature for the synthesis of guanidines.

In an effort to efficiently synthesise a diverse variety of *N*,*N*′,*N*″-trisubstituted guanidines **48**, an effective phase-transfer-catalysed alkylation of protected guanidines **49** was unveiled (Scheme 21) [40]. A variety of amines **50** were reacted with *S*-methyl-*N*,*N*′-bis-Boc-isothiourea **45** in the presence of $HgCl_2$, as this is known in the literature to be very efficient in the preparation of protected guanidines.

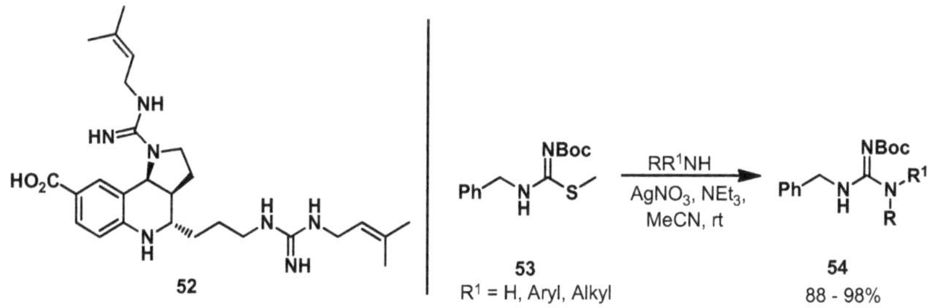


**Scheme 22**

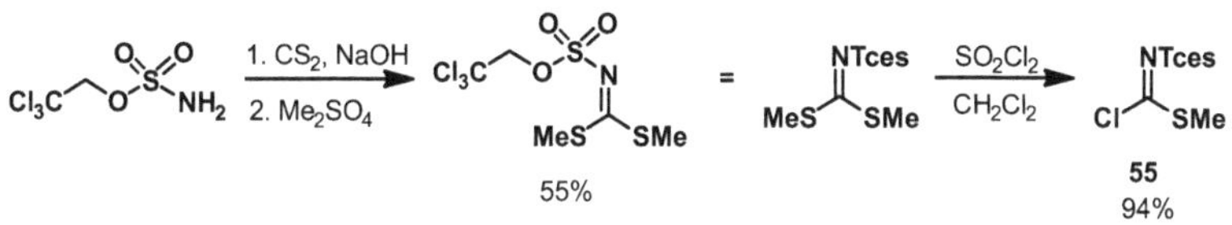


**Scheme 23**

**Scheme 24**

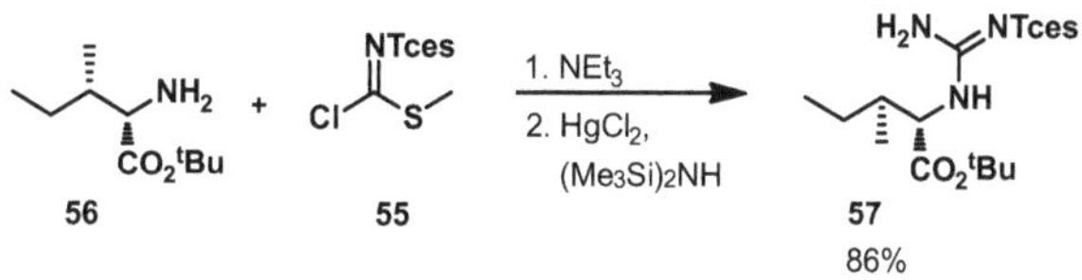


Then, in a biphasic system of $CH_2Cl_2$ and $H_2O$, the compounds **49** were reacted with KOH, a phase-transfer catalyst ($Bu_4NI$, 0.1 equiv.) and a variety of alkylating agents **51** (Scheme 21). This process allowed the rapid formation of highly diversified guanidines **48** in good to excellent yields. This reactivity, however, was only reliable when both R and $R^1$ were not equal to H, as this led to non-specific alkylation at either nitrogen or, in some cases, to bis-alkylation. In Ma's synthesis of martinellic acid **52**, a novel method of guanidylation was devised [41]. It was found that $AgNO_3$ is able to promote the guanidylation of *S*-methylisothiourea derivatives **53** with a variety of amines to form guanidine products **54** in excellent yield (Scheme 22).

The reaction was shown to work well with aliphatic amines, sterically hindered secondary amines and also with aniline. This new method was formulated because hindered secondary amines are often difficult to guanidylate and initial experiments using typical guanidylation methodologies proved unfruitful. The method allowed the team to install the key guanidine functional groups with relative ease.

An interesting guanidylating agent designed by Du Bois and used in his synthesis of guanidine natural products is Tces (2,2,2-trichloroethoxysulfonyl)-protected imidochloride **55** [42]. This compound can be prepared as shown in Scheme 23.

The reagent **55**, although slightly laborious to prepare, has shown to be an effective way of introducing guanidines (Scheme 24), which are to be used in

Scheme 25

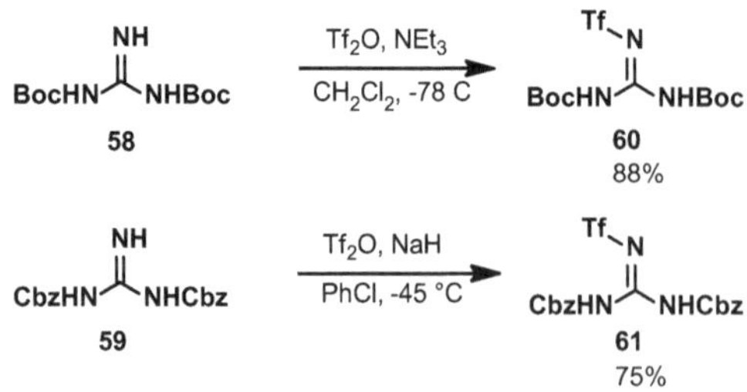


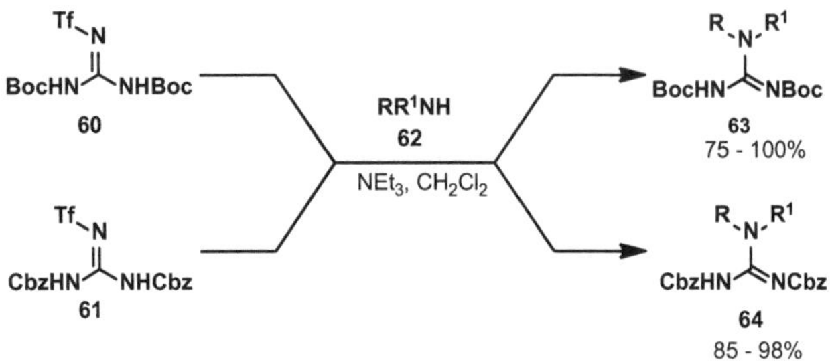


Scheme 26

C–H amination reactions [43]. Romo has also made use of this reagent in his synthesis of the natural product phakellin [44].

## 2.6 N,N′,N″-*Trisubstituted Guanidines as Guanidylating Agents*

Since 1998, it has been known that guanidines can be synthesised from guanidine precursors by coaxing one of the nitrogens into being an effective leaving group. This can be done effectively as discovered by Goodman and co-workers when using triflate-derivatised guanidines (Scheme 25) [45].

Firstly, either *N,N′*-bis-Boc-guanidine **58** or *N,N′*-bis-Cbz-guanidine **59** is synthesised; subsequent exposure to triflic anhydride introduces a trifluoromethanesulfonyl group to the remaining unsubstituted nitrogen atom of the guanidine moiety (Scheme 25, **60**, **61**). The two carbamate protecting groups (either Boc or Cbz) pull electron density away from the central C atom of the guanidine. Nucleophilic attack can follow, and the triflated amine can act as a leaving group (Scheme 26). The relative stability (due to the delocalisation of the ensuing negative charge) of this expelled group is what drives this reaction forward and allows these reagents to be effective as guanidylating agents. Sterically hindered aliphatic amines and primary amines react particularly well under these conditions; however, limited studies have been conducted on the efficacy of these reagents with electron-deficient anilines [46].

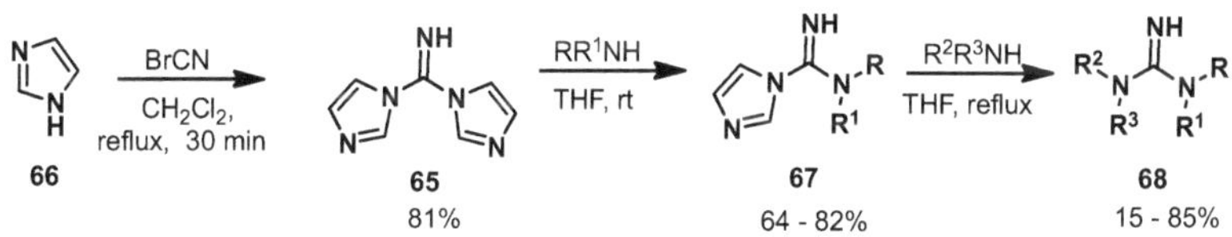


Scheme 27

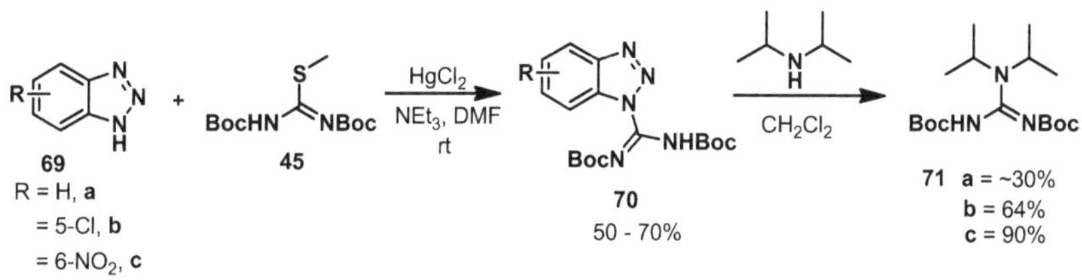


Scheme 28

The use of di(imidazol-1-yl)methanimine **65** as an effective guanidylating agent has also been shown to be synthetically useful in the preparation of derivatised guanidines [47]. By exposing imidazole **66** (0.5 equiv.) to cyanogen bromide (BrCN, 1.0 equiv.), **65** can be synthesised in good yield (Scheme 27, **65**: 81%).

The presence of two imidazole leaving groups facilitates initial displacement of a single imidazole by an amine nucleophile. The products **67** can then be isolated and exposed to a further equivalent of another amine, generating structurally diverse guanidines **68** in acceptable yields (Scheme 27). The success of these reactions is dependent on the nucleophilicity of the attacking amines; electron-deficient anilines (such as 4-nitroaniline) did not react in this case [47]. Similar pyrazole derivatives have also been designed and shown to be effective as guanidylating agents [48]. A further study showed the effects of electron-withdrawing substituents on the pyrazole ring in forming guanidines [49].

Subsequent research in this area investigated the use of benzotriazoles **69a** as leaving groups in guanidylation reactions [50]. A comparative study was conducted on the influence of decreasing the electron density of the benzotriazole group using chloro **69b** and nitro **69c** groups (Scheme 28). Increasing the electrophilic character and leaving group ability of the benzotriazoles, by allowing them to facilitate the resulting negative charge arising from expulsion, increased their reactivity towards amine nucleophiles.

## 2.7 *Urea Derivatives in Guanidine Formation*

As an oxygen analogue of guanidine, urea would seem like an obvious precursor in the synthesis of guanidines. However, urea has received much less attention in comparison to its sulphur analogue, thiourea, presumably due to its stable nature.

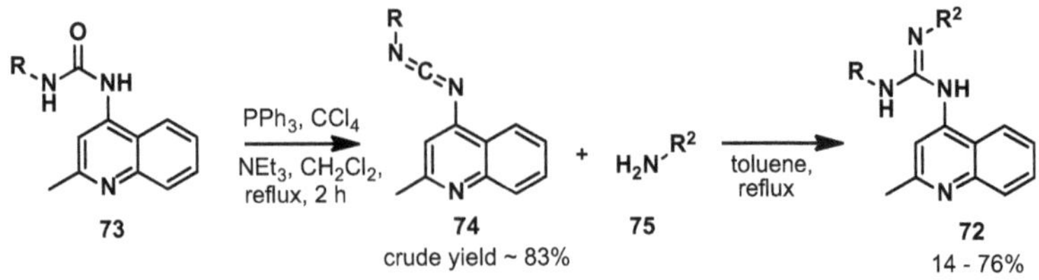


Scheme 29

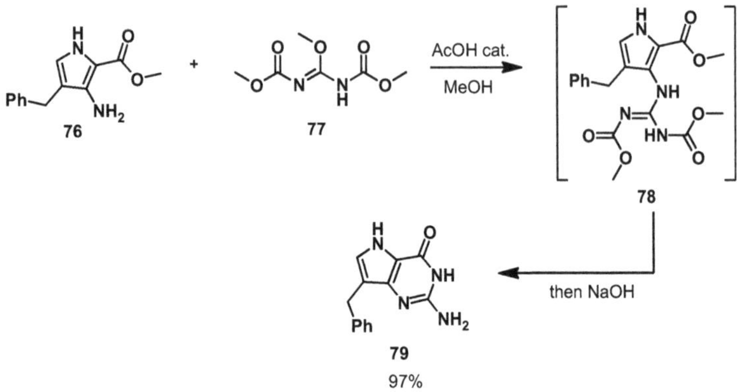


Scheme 30

The literature examples are typically either low yielding or have a low substrate scope [51]. One such example is the synthesis of a library of quinoline guanidine derivatives **72** in the hope of discovering novel anti-inflammatory agents [52]. Thiourea-derivatised quinolines **73** were exposed to $PPh_3$, $CCl_4$ and $NEt_3$ under reflux in $CH_2Cl_2$ using Appel-type conditions [53] to form crude carbodiimides **74** which, without purification, were exposed to a number of amines **75** in different solvents yielding *N,N′,N″*-trisubstituted guanidines **72** in varying yields (Scheme 29). The inability to purify the intermediate carbodiimide is an obvious drawback. This reaction gives hugely varying yields, with some substrates giving yields of just 14%.

Aiming to synthesise 7-substituted pyrrolo[3,2-*d*]pyrimidines, 3-aminopyrrole derivative **76** was reacted with bis-protected pseudourea **77** in the presence of catalytic amounts of AcOH in MeOH (Scheme 30) [54]. This led to guanidylated intermediate **78** which, upon exposure to NaOH, was deprotected and cyclised to form the desired product **79**. Initial attempts to guanidylate amine **76** using a thiourea analogue of **77** also led to the desired product. However, as it is typical with the reaction of aromatic amines and thiourea derivatives, $HgCl_2$ was required to promote the reaction. The authors decided that, as the compounds were to be used in biological testing, the presence of ppm amounts of mercury was unacceptable and that the use of urea precursor **77** was a much safer alternative.

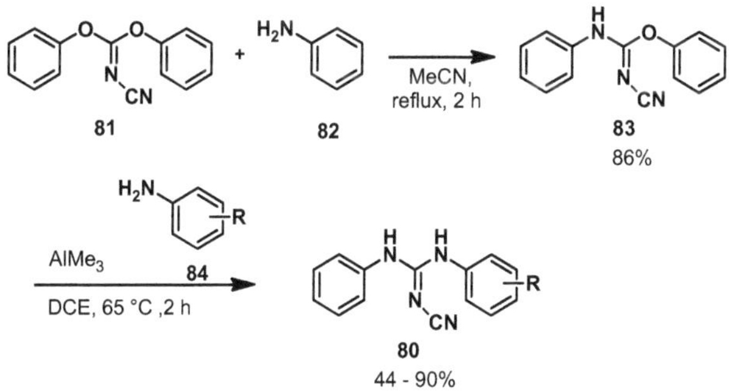


Scheme 31

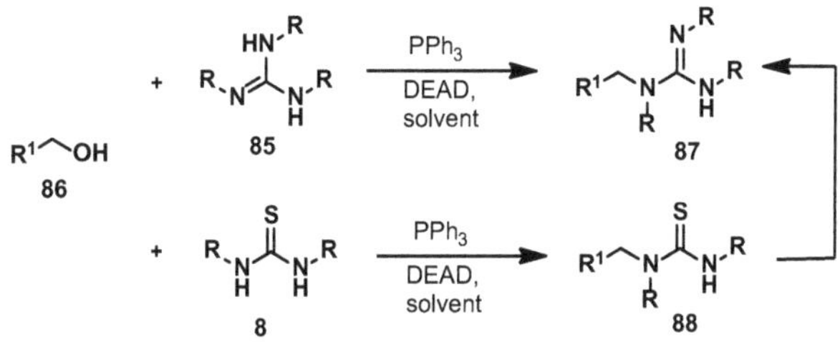


Scheme 32

In an effort to synthesise cyanoguanidine derivatives **80**, a novel method of guanidylation was discovered by Atwal and co-workers at BMS, Princeton [55]. *N*-Cyano-*O*-phenylisourea **81** was found to react favourably with nonbasic amines such as aniline **82** to form derivatised ureas **83** that were transformed into cyanoguanidines **80** when $AlMe_3$ was used as a promoting agent (Scheme 31).

High yields were obtained when electron-rich anilines **84** such as 4-methoxyaniline were used (90%), whilst electron-deficient anilines such as 4-nitroaniline gave no reaction.

## 2.8 *The Mitsunobu Reaction to Form Guanidines*

The utility of hydroxyl groups as synthetic handles has been well established throughout chemical history. The ability to convert hydroxyl groups into varying substituents by means of the Mitsunobu reaction has proven to be ubiquitous in synthesis since its discovery in 1967 [56] as is discussed in the numerous reviews on the topic [57]. As a means of guanidine introduction, the ability to replace primary hydroxyl functionalities is particularly useful when it is advantageous to have hydroxy precursors as opposed to aliphatic amines.

Both protected guanidines **85** and thiourea guanidine precursors **8** can be reacted with primary alcohols **86** (Scheme 32) in the presence of triphenylphosphine ($PPh_3$)

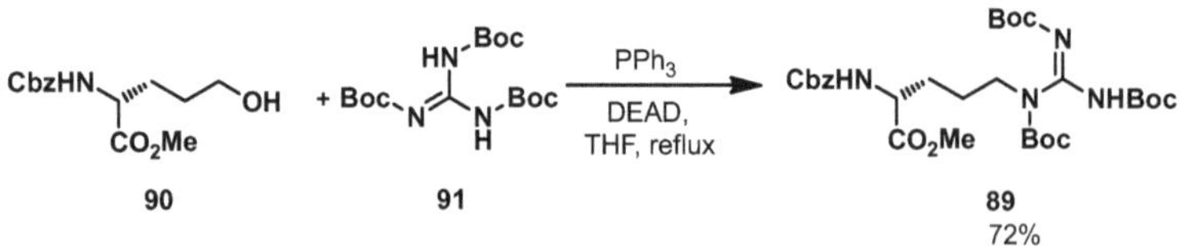


**Scheme 33**

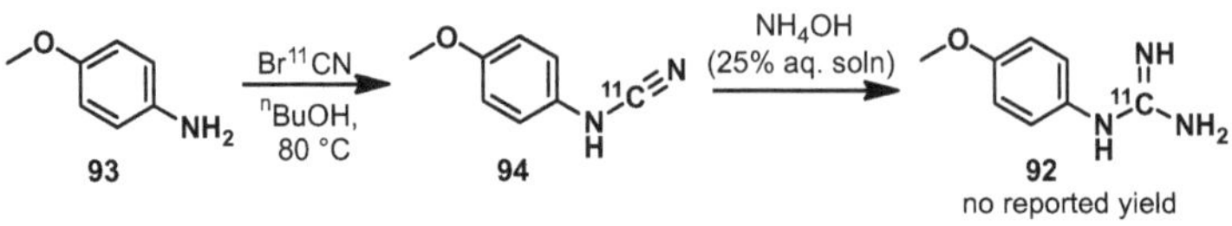


**Scheme 34**

and diethylazodicarboxylate (DEAD) to form alkylated derivatives **87** and **88** (Scheme 32) [58].

This methodology has been applied to the preparation of protected arginine derivatives [59] **89** (Scheme 33) and natural products [60] in acceptable yields.

## 2.9 *Cyanamides in Guanidine Synthesis*

A rather dated method for the synthesis of guanidines makes use of the formation of cyanamides and subsequent reaction with derivatised amines. Functionalised anilines can be reacted with cyanamide to form aryl guanidines [61], and, analogously, phenyl cyanamides can be synthesised and then further reacted with a variety of amines, also forming aryl guanidines [61].

The synthesis of $^{11}$C radiolabelled aryl guanidines **92** is of great importance as guanidines are involved in numerous biological activities, and labelling them with short-lived positron emitters ($^{11}$C $t_{1/2} = 20.3$ min) such as $^{11}$C would allow their use in both in vivo positron emission tomography (PET) studies and in vitro assays [61]. Thus, the preparation of $^{11}$C-labelled aryl guanidines was achieved using cyanamide chemistry (Scheme 34). Firstly, the appropriate aniline **93** was reacted with $^{11}$C-labelled cyanogen bromide, and then this aryl cyanamide **94** was exposed to ammonia in different solvents to afford the desired products **92** [61].

In a report by Kim and co-workers [62], this reactivity was utilised in the synthesis of aryl guanidines which were required in the synthesis of inhibitors of the NF-κB transcription regulation related to TNF-α cytokine release. A number of

Scheme 35

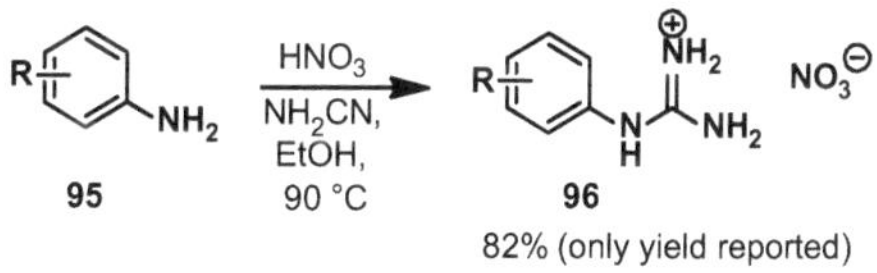

Scheme 36

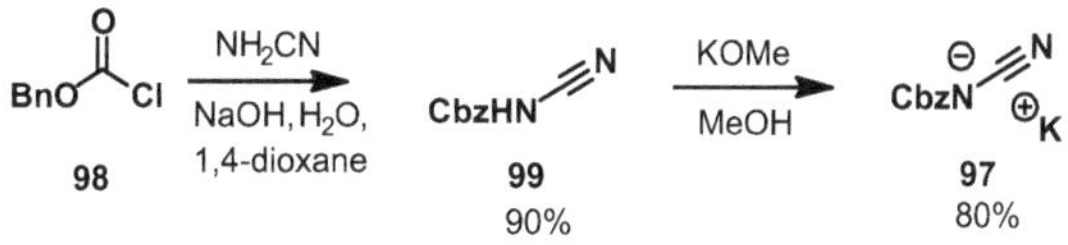

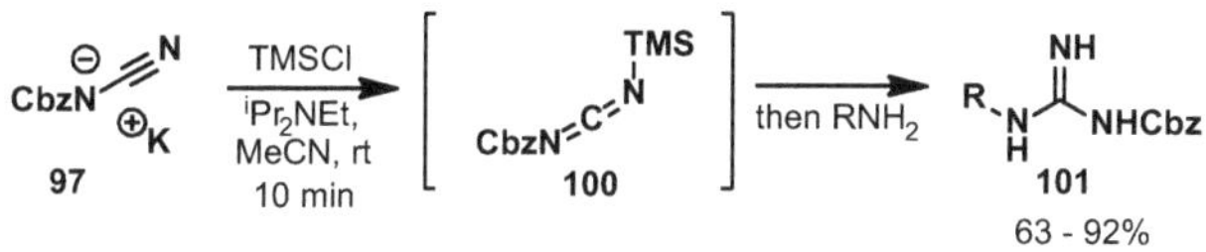

Scheme 37

Scheme 38

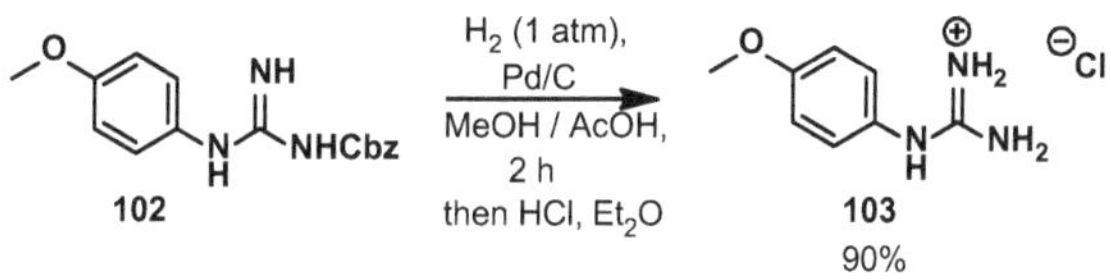

anilines **95** were reacted with conc. nitric acid and cyanamide in ethanol at 90°C to afford aryl guanidines **96** (Scheme 35).

A more efficient method for the reaction of anilines and also aliphatic amines with cyanamides has recently been unveiled by Looper and co-workers at the University of Utah [63]. Derivatised cyanamide **97** was first synthesised by reacting cyanamide with benzyl chloroformate **98** forming the Cbz-protected cyanamide **99** (Scheme 36).

This was then converted into the potassium salt **97** by exposure to potassium methoxide in methanol. Compound **97** could then be activated in the presence of TMS-Cl and a base, forming a carbodiimide reactive intermediate **100** which can be nucleophilically attacked by a number of amines to form mono Cbz-protected guanidines **101** (Scheme 37).

Aliphatic nucleophilic amines gave the highest yields with short reaction times (15 min), whilst both hindered and non-nucleophilic substrates such as anilines required longer reaction times and gave lower yields. The Cbz group (in compounds such as **102**) could then potentially be removed by Pd-catalysed hydrogenation (Scheme 38) to afford guanidine **103**.

## 2.10 *Copper-Catalysed Cross-Coupling Chemistry in Guanidine Synthesis*

Copper coupling chemistry has seen a renaissance in the past decade [64]. The harsh conditions historically required for Ullmann reactions have been discarded and replaced by mild, copper-catalysed reactivity [65]. The ability to form C–N bonds in a facile manner is of clear importance in the synthesis of guanidines, and copper coupling chemistry allows a type of *umpolung* reactivity in comparison to the typical methods of guanidine synthesis. As opposed to nucleophilic anilines reacting with a guanidine precursor, an aryl halide is exposed to a nucleophilic guanidine source. This reactivity allows for facile assembly of aryl guanidines.

As it is typical with copper coupling chemistry, the exact mechanism of the catalytic cycle is still unclear. However, a possible catalytic cycle that would be in agreement with previous findings in copper chemistry has been suggested by Ma and co-workers at Shanghai Institute of Organic Chemistry (SIOC) [66]. It is known that α-amino acids and CuI can form chelate complexes such as **104** (Fig. 3). The formation of the Cu(I)–amino acid species makes it more reactive towards oxidative addition and may also stabilise the oxidative addition intermediates, thereby promoting the coupling reaction [67]. Complex **104** may then coordinate to guanidine to give complex **105**, whose oxidative addition to an aryl halide could afford Cu(III) complex **106**. The presence of a base in the catalytic cycle could then result in the formation of **107** which upon reductive elimination could generate *mono*-arylated guanidines and regenerate the Cu(I) species **104**.

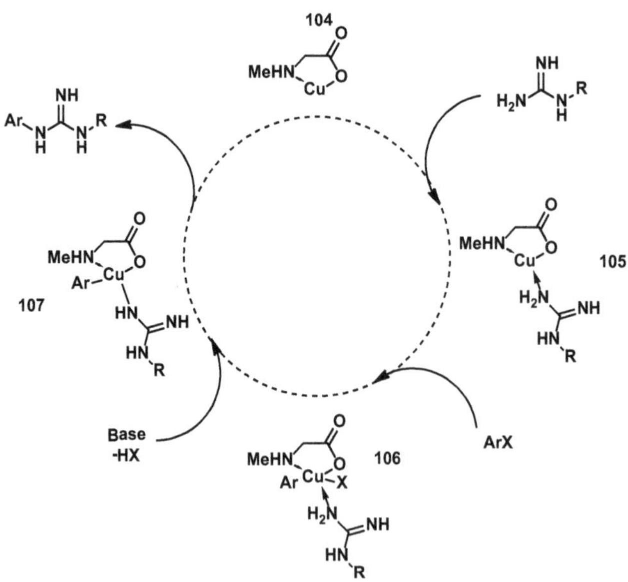


**Fig. 3** Proposed catalytic cycle for copper-catalysed guanidylation [67]

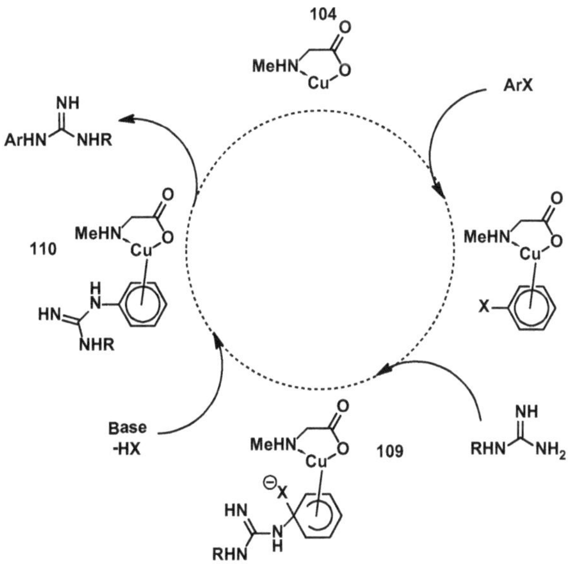


**Fig. 4** Alternative catalytic cycle for copper-catalysed guanidylation [68]

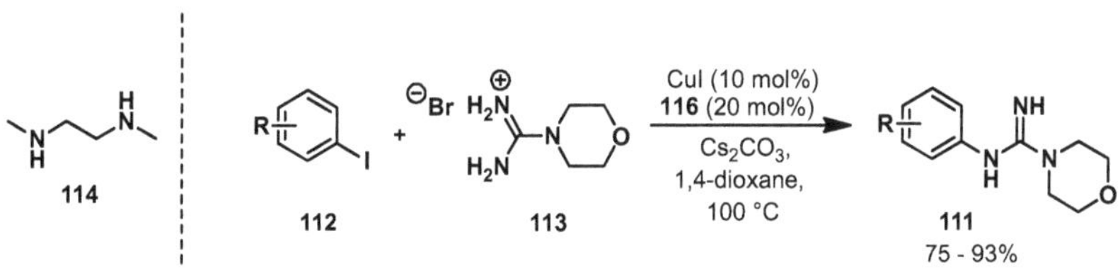


**Scheme 39**

Another plausible mechanism for copper-catalysed coupling chemistry is depicted in Fig. 4 and has been previously suggested [66, 68]. As in the previous mechanism, a Cu(I) complex will be formed by coordination to the α-amino acid ligand. Chelated Cu(I) may possibly coordinate to an aryl halide forming the π-complex **108** which would be electron deficient and more susceptible to nucleophilic attack by an amine giving **109** (Fig. 4). Subsequent expulsion of the halide leaving group to form **110** and dissociation of the copper complex regenerate the active Cu(I) species and yield the desired product. Although the actual mechanism is not exactly known, both have been often used to explain results arising from copper coupling chemistry.

In an investigation into the synthesis of benzimidazoles, the formation of aryl guanidines **111** by means of copper-catalysed cross-coupling chemistry was discovered (Scheme 39) [69]. Aryl iodides **112** and guanidines **113** could be coupled together in the presence of CuI and ligand **114** to yield aryl guanidines **111**. This

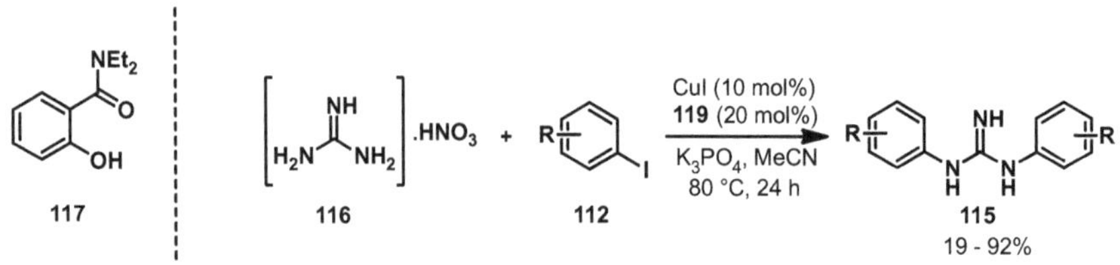

Scheme 40

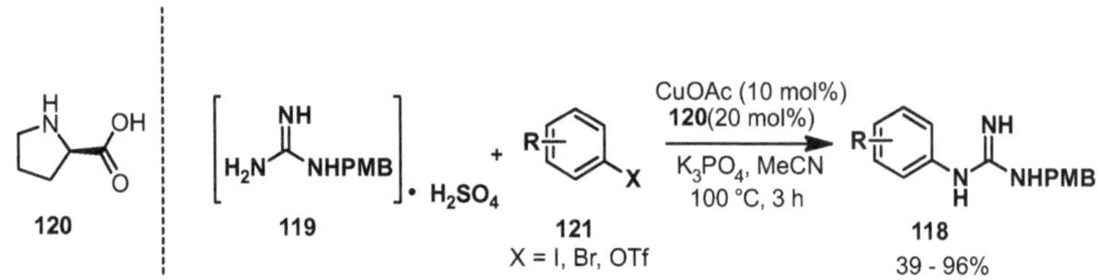

Scheme 41

was the first report of its kind (with six examples) and was followed by investigations from other research groups developing the result further.

In 2010, a report from Antilla and co-workers described the formation of *N*, *N′*-disubstituted aryl guanidines **115** [70]. Guanidinium nitrate **116** was chosen as the source of nucleophilic guanidine, and a wide variety of aryl iodides **112** were selected as the coupling partners (Scheme 40).

Thus, in the presence of catalytic amounts of CuI (10 mol%), ligand **117** (*N*, *N′*-diethylsalicylamide, 20 mol%) and $K_3PO_4$ as base in acetonitrile at 80°C for 24 h, successful couplings between the two substrates were achieved with varying yields. However, *mono*-arylated products could not be obtained, with diarylation being observed in all cases (Scheme 40).

A 2012 report also investigated the feasibility of forming C–N bonds through copper catalysis with the aim of preparing *mono*-arylated guanidines **118** (Scheme 41) [71]. A number of guanidine sources were investigated using both copper and palladium catalysts. Protected guanidines were chosen as the nucleophilic substrates due to their lower basicity and less reactive nature than guanidine itself. Palladium catalysis proved to be ineffective; however, initial experimentation with copper catalysts proved to be promising with a high ratio of *mono*-arylated to diarylated products being observed. *p*-Methoxybenzyl (PMB) guanidine **119** was chosen as the most suitable form of protected guanidine. PMB-protected guanidines were made from benzylamine and 2-methylthiopseudourea in 62–80% yields.

A range of α-amino acid ligands were investigated for their efficacy in this cross-coupling reaction, as were varying copper sources, bases, solvents, temperatures and times. Optimal results were observed when CuOAc (10 mol%), ligand **120** (20 mol%) and $K_3PO_4$ in MeCN at 100°C were employed for 3 h. A wide variety of

Scheme 42

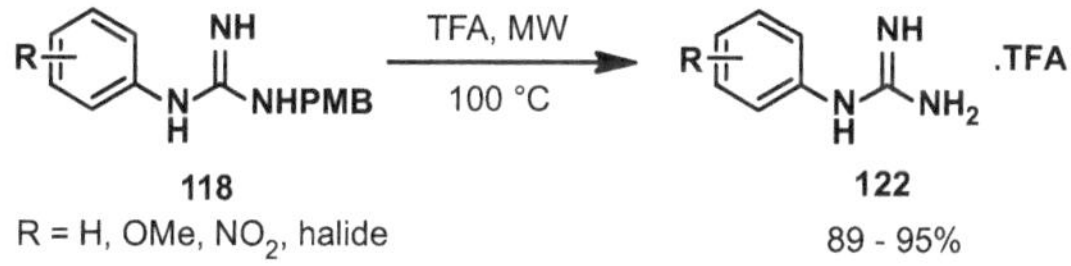


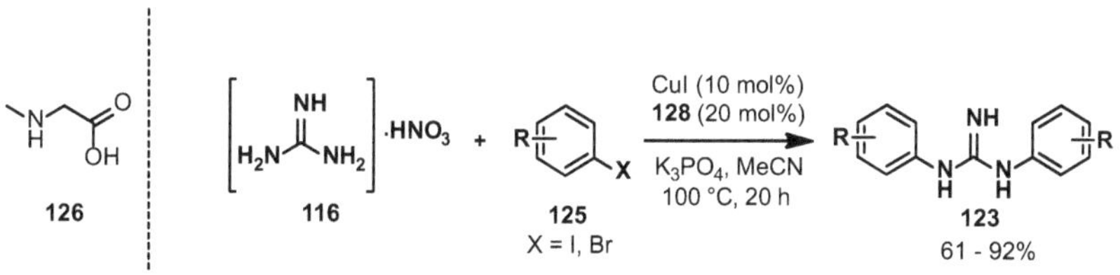


Scheme 43

aryl halides **121** were coupled in good to excellent yields (Scheme 41). This is one of the few literature examples of C–N bond formation utilising copper coupling chemistry with an aryl triflate as a coupling partner [72].

Deprotection of the PMB-protected guanidines **118** was then accomplished by microwave irradiation in TFA at 100°C (Scheme 42). This afforded the corresponding guanidines as trifluoroacetate salts **122** in excellent yields (Scheme 42, 89–95%). Due to the harsh nature of these deprotection conditions, only a few functional groups as substituents on the aryl ring were investigated.

In an attempt to form both symmetrical and unsymmetrical *N*,*N*′-diaryl guanidines (**123** and **124**, respectively), Ma and co-workers investigated the use of copper cross-coupling chemistry [73]. As with previous discoveries [70, 71], initial investigations into the coupling of aryl halides with guanidines were concerned with optimising the reaction conditions to afford symmetrical *N*,*N*′-diaryl guanidines. Hence, aryl halides **125** and guanidinium nitrate **116** were successfully coupled using a copper(I) source, an α-amino acid ligand **126**, base and solvent. The optimal reaction conditions are shown in Scheme 43.

As has been previously reported [67, 69, 70], the use of CuI (10 mol%) and $K_3PO_4$ in MeCN was optimal, with *N*-methylglycine (**126**, 20 mol%) being the ligand of choice. Under these conditions, a number of aryl iodides could be diarylated in good yields, and aryl bromides were also successfully coupled with guanidinium nitrate in acceptable yields.

The poor ability of electron-deficient aryl halides **127** to form diarylated products indicated the potential for forming unsymmetrical *N*,*N*′-diaryl guanidines in a sequential one pot process. Initial investigations coupled 4-nitroiodobenzene with guanidinium nitrate **116** in conditions analogous to those reported for the previous diarylation [70]. After 10 h, an electron-rich aryl halide **128** (e.g. 4-methoxyiodobenzene) was added to the reaction mixture and allowed to react for a further 8 h (Scheme 44). As desired, the reaction resulted in the formation of unsymmetrical *N*,*N*′-diaryl guanidines **124** in reasonable yields.

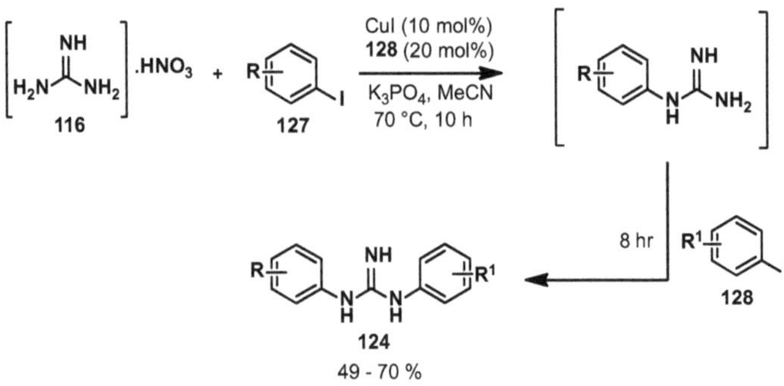


Scheme 44

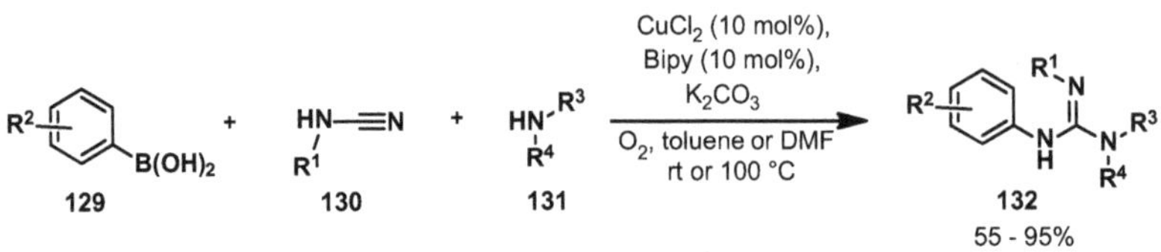


Scheme 45

In 2013, a three-component copper-catalysed coupling reaction between arylboronic acids **129**, cyanamides **130** and amines **131** was reported [74]. This reaction allows access to trisubstituted aryl guanidines **132**, with a high potential for introducing diversity (Scheme 45).

This reaction is thought to proceed through transmetallation of a copper (II) species generated by oxidation of the Cu(I) salt in the presence of $O_2$. Coordination of the cyanamide and deprotonation, followed by tautomerisation, would ensure the generation of the highly electrophilic carbodiimide. This can then be attacked by the amine, and subsequent oxidation of the Cu(II) species to Cu(III) enables reductive elimination, generating the guanidine product and regenerating the Cu(I) species. This possible catalytic cycle, as described by the authors, is presented in Fig. 5.

In an effort to synthesise derivatised guanidines, transition metal-catalysed allylic substitution of *N*-Boc-protected guanidines was investigated (Scheme 46) [75]. Miyabe et al. reported that both mono and double allylic substitution could occur. Guanidine bearing two electron-withdrawing groups could act as a nucleophile in an allylation reaction to form the mono-allylated product, whilst tri-Boc-guanidine when exposed to similar conditions could afford the diallylated products. The regiocontrol in the allylic substitution of unsymmetrical allylic substrates was investigated by using both Pd and Ir catalysis.

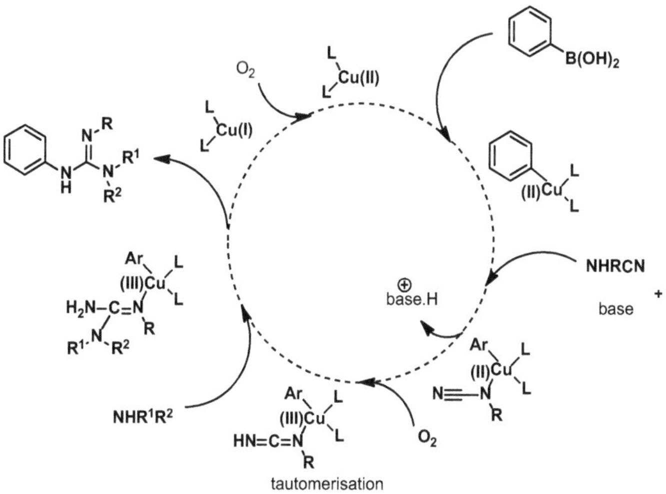


**Fig. 5** Proposed catalytic cycle for copper-catalysed three-component coupling [74]

**Scheme 46**

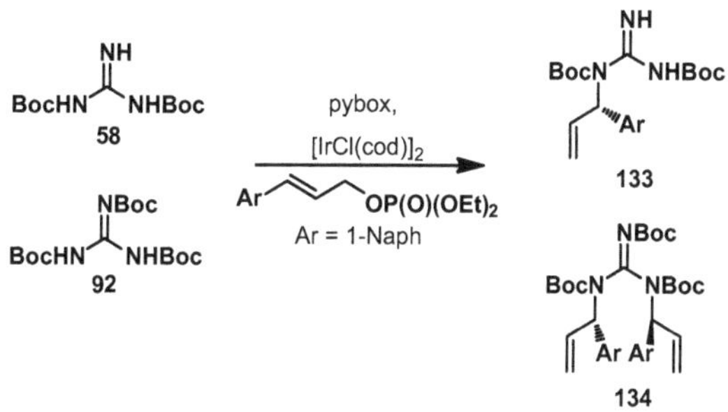


# 3 Synthesis of Cyclic Guanidines

The ubiquitous nature of cyclic guanidine moieties throughout natural products [16] and compounds of medicinal interest [15] has ensured the development of a wealth of methodologies designed for their facile synthesis. Herein we present some selected examples of literature methods for the formation of five-, six- and seven-membered rings incorporating the guanidine motif in their structure.

## 3.1 *Five-Membered Rings*

Amongst the most facile method for the synthesis of five-membered rings containing guanidines is the use of *N,N′*-disubstituted-2-imidazolidinethione **135**

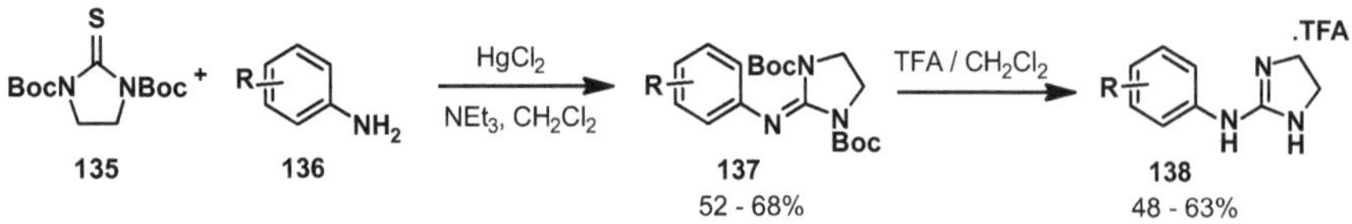


Scheme 47

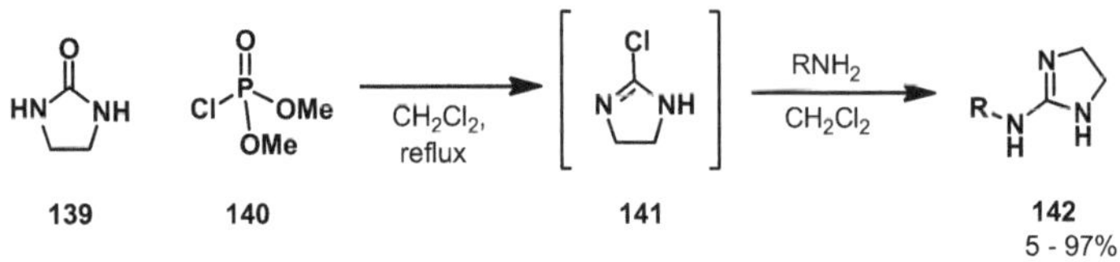


Scheme 48

as a five-membered cyclic analogue of thiourea. In this method, **135** can be reacted with a variety of amines **136** in the presence of a Lewis acid promoter such as $HgCl_2$ (Scheme 47) to form protected guanidines **137** [76]. Upon deprotection, a wide variety of *N*-substituted-2-aminoimidazolines **138** can be generated as has been shown by Rozas and co-workers in their development of $\alpha_2$-adrenoceptor ligands [77].

The oxygen analogue of **135** has also shown to be of some synthetic utility in this type of reactions, and no intermediates are isolated in this one-pot procedure [78]. Imidazolidin-2-one **139** is exposed to dimethyl chlorophosphate **140** forming intermediate **141** in situ, which in the presence of an amine will react to form the desired *N*-substituted imidazolines **142**, displacing the halide leaving group (Scheme 48). An extensive study has not been carried out on this type of reactivity, with only few examples known. This methodology has also been applied to the synthesis of *N*-substituted-2-amino-1,4,5,6-tetrahydropyrimidines [78].

Perhaps the most atom-economical method for the generation of five-membered cyclic guanidines is the exposure of 1,2-diamines to cyanogen bromide. Initial reaction of an amine such as compound **143** with cyanogen bromide will form compound **144**, in which the cyanamide functionality is now ideally located for a 5-*exo*-dig cyclisation to form the desired product **145**, as demonstrated in the polymer-supported chemistry shown in Scheme 49. Using this very simple chemistry, a vast library of compounds could be synthesised ($R^1$ (26) × $R^2$ (26) × $R^3$ (26) × $R^4$ (42) = a total of 738,192 compounds) [79].

The synthesis of bicyclic guanidine-containing compounds such as 2,3,5,6-tetrahydro-1*H*-imidazo[1,2-a]imidazole **146** has received considerable attention due to their unique properties as superbases and application in organocatalysis. Initial investigations into their synthesis began with undesirably lengthy undesirable syntheses [80]. However, in a patent of 1990, an efficient synthesis from linear triamines **147** was devised by exposing them to $CS_2$ in *p*-xylene (Scheme 50) to generate compound **148** which then cyclises to afford thiourea **149** [81].

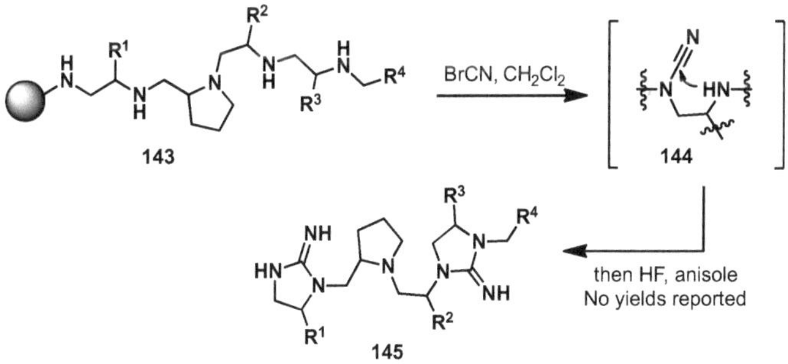


Scheme 49

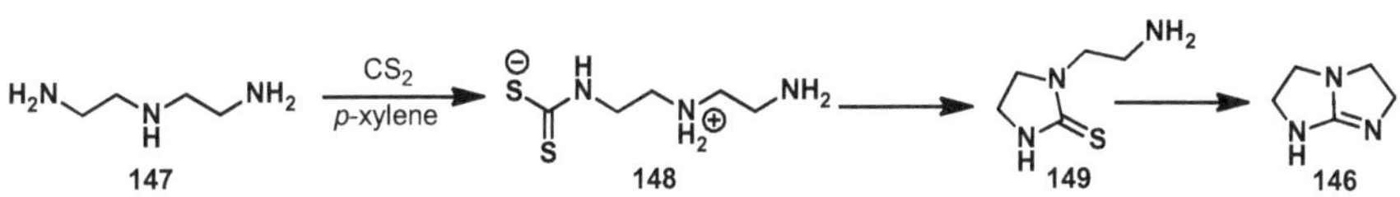


Scheme 50

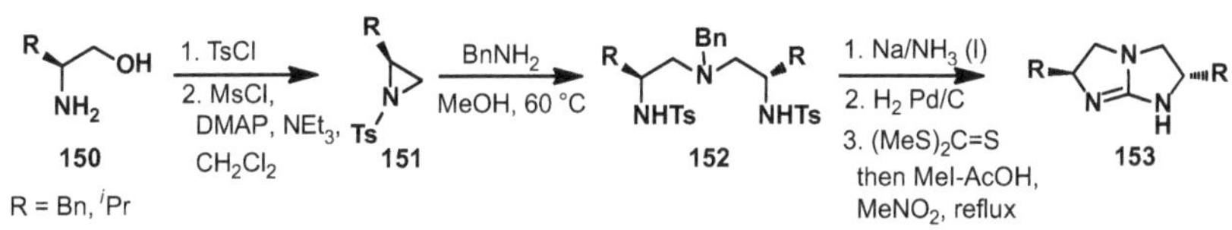


Scheme 51

A more elegant synthesis for derivatised bicyclic guanidines results in chiral products [82]. Initial tosylation of the amino group of a homochiral 1,2-amino alcohol **150**, followed by mesylation of the hydroxyl group, generates aziridine **151**. Exposure to benzylamine (0.5 equiv.) stereospecifically opens aziridine **151** followed by reaction with another molecule of **151** to generate triamine **152**. *N*-Deprotection and treatment of **152** with dimethyl trithiocarbonate, followed by methylation and heating, generate the desired bicyclic guanidines **153** (Scheme 51).

Corey also designed a synthesis of bicyclic guanidine compounds whilst investigating enantioselective Strecker reactions catalysed by chiral bicyclic guanidines; however, his synthetic route involved nine steps [83].

α-Chloroaldoxime-*O*-methanesulfonates, such as **154**, are known to undergo Tiemann rearrangement [84] (the aza analogue of the Lossen rearrangement [85]) in the presence of nucleophilic amines. The resultant carbodiimide **155** is then available for nucleophilic attack by an external amine to generate both derivatised guanidines and imidazolines **156** (Scheme 52). This method was ingeniously used by Yamamoto et al. in their synthesis of a variety of guanidine-containing compounds [84]. A representative example is shown in Scheme 52.

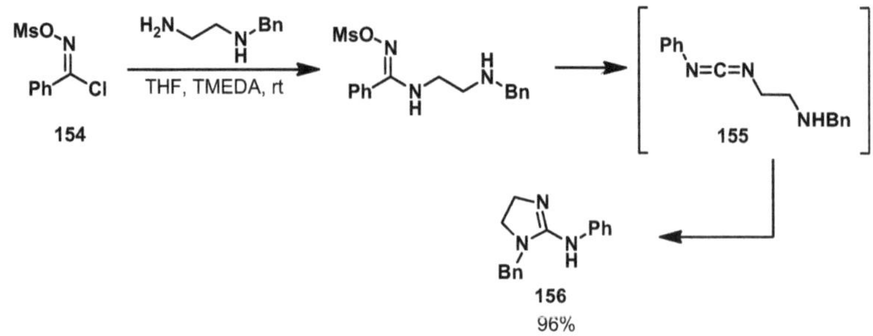

Scheme 52

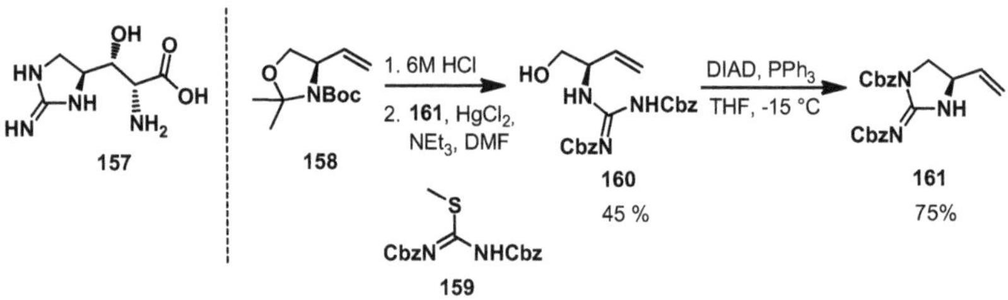

Scheme 53

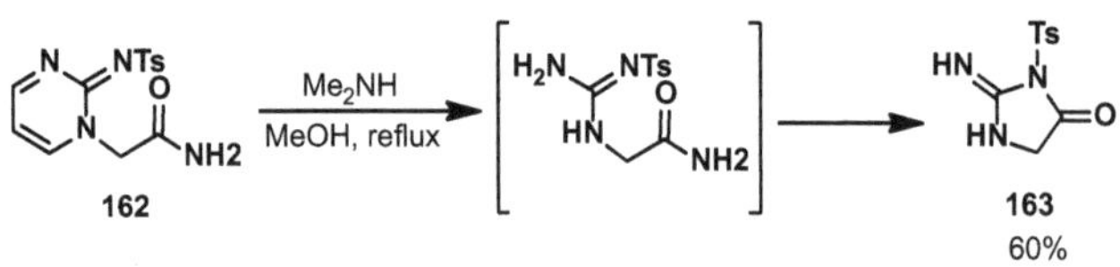

Scheme 54

As part of an ongoing programme into the synthesis of the antibiotic mannopeptimycin β, van Nieuwenhze and co-workers applied a Mitsunobu reaction in their synthesis of cyclic guanidine intermediate **157** [60]. The synthesis began with deprotection of protected amino alcohol **158** followed by guanidylation using thiourea derivative **159** to afford compound **160**. A subsequent Mitsunobu reaction closed the five-membered ring in good yield, forming the desired cyclic guanidine **161** (Scheme 53).

Interestingly, there was no aziridine side-product reported in the conversion of **160** into **161** even though the unprotected amine of the guanidine would presumably be more nucleophilic than the carbamate nitrogen. Presumably, the more favourable formation of a five-membered ring dominates the reactivity of this species.

In an interesting example of ring-opening/ring-closing reactions, a report in 2007 described the conversion of 2-aminopyrimidine derivatives **162** into cyclic guanidine moieties **163** as shown in Scheme 54 [86].

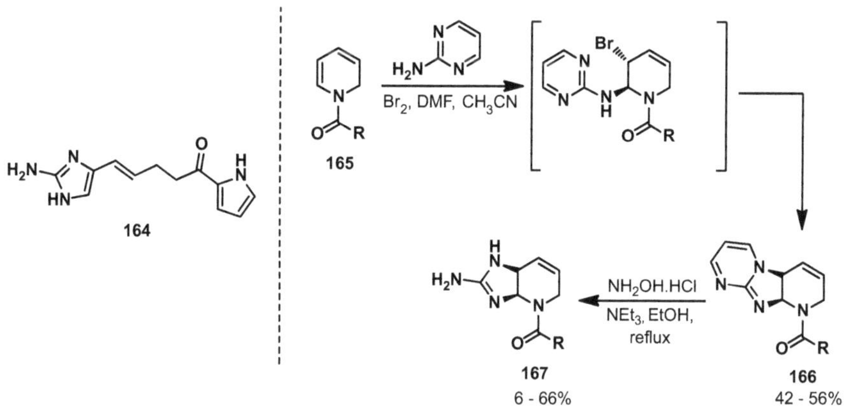


**Scheme 55**

Ensuring tosyl protection of the exocyclic amino group of 2-aminopyrimidine facilitates alkylation of a ring nitrogen of the pyrimidine affording **162**. Compounds of this nature are known to undergo ring opening and cleavage of the carbon framework connecting the two nitrogens of the pyrimidine ring [86]. Exposure of **162** to $MeNH_2$ and heating results in ring cleavage, and then condensation of the guanidine functionality on to the amide results in an unusual route to cyclic guanidines **163**.

Similar reactivity has been used effectively by Al-Mourabit in the synthesis of clathrodine **164** (Scheme 55) [87]. Initial investigations into the addition of a protected guanidine to an alkene **165** exposed to bromine did not result in guanidine addition to the olefin.

However, when the protected guanidine was replaced with 2-aminopyrimidine, initial displacement of a bromine atom and resulting ring closure to form **166** occurred. Treatment of **166** with $NH_2OH$.HCl in refluxing EtOH resulted in ring cleavage and cyclic guanidine **167** formation in good yield (Scheme 55).

In 2011, a novel one-pot synthesis of cyclic guanidines was devised allowing for the formation of a diverse array of guanidine products **168** [88]. A variety of alkenes **169** were exposed to NBS, amines **170** and cyanamides resulting in the desired cyclic compounds (Scheme 56).

Initial activation of the olefin, followed by nucleophilic attack by the cyanamide, generates a highly electrophilic species **171**. This in turn undergoes intramolecular attack by the amino group, generating the desired five-membered ring **168**.

One of the simplest approaches to the preparation of five-membered cyclic guanidine containing molecules is a reaction pathway prevalent throughout nature, the biosynthesis of creatinine **172** (Fig. 6).

This biosynthetic process involves the intramolecular condensation of the nucleophilic guanidine functionality of arginine onto its carboxylate to form a five-membered ring as is present in creatinine [89], and this pathway can be easily replicated in the laboratory as displayed in Bazureau's synthesis of dispacamide A

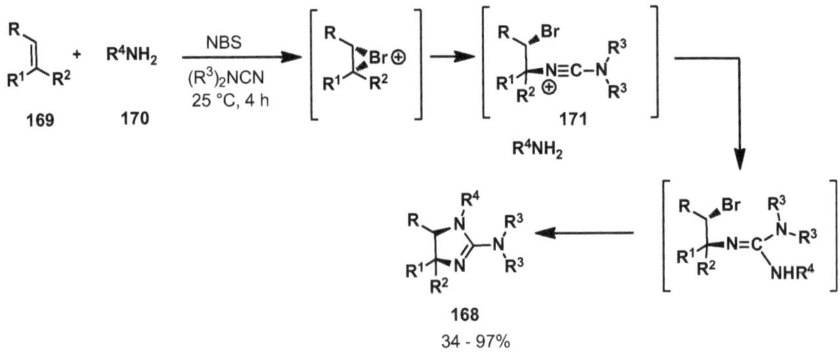


Scheme 56

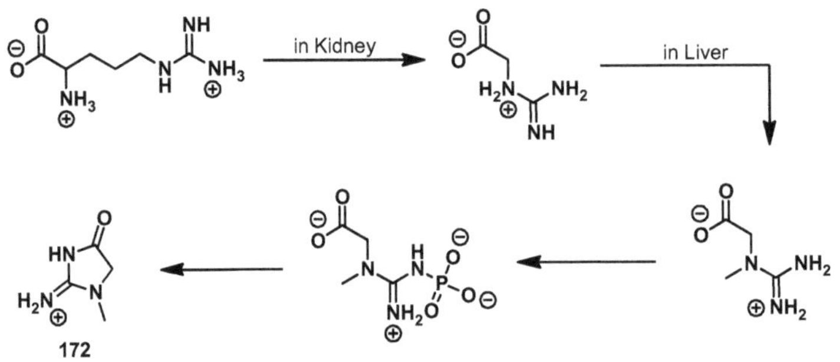


**Fig. 6** Biosynthesis of creatinine

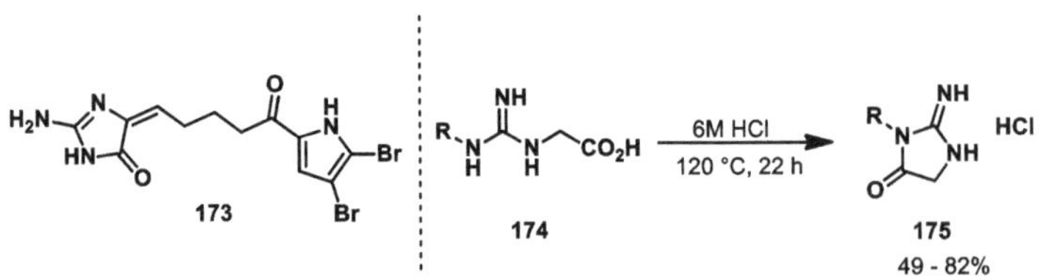


Scheme 57

**173** (Scheme 57) [90]. Exposure of guanidine **174** to acidic conditions caused ring closure to generate cyclic guanidine **175**.

In Danishefsky's efforts towards the total synthesis of spiroleucettadine **176**, a similar type of reactivity was employed [91]. Initial guanidylation of **177** utilising compound **178** as a guanidylating agent, followed by ring opening of the lactone and then ring closing of the guanidine functionality, resulted in a ring-opened isomeric form **179** of the reported structure of **176** (Scheme 58).

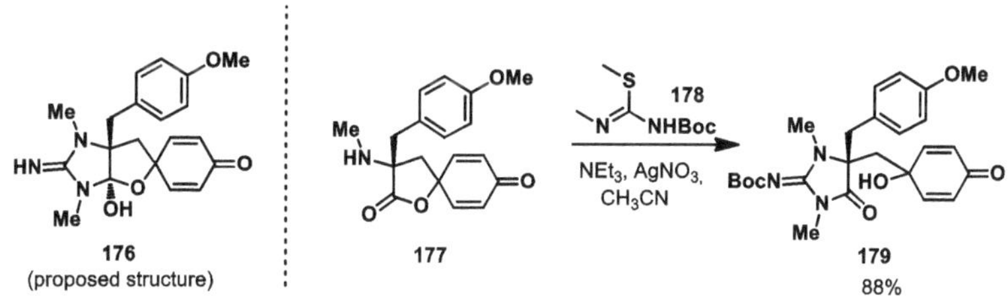


Scheme 58

Scheme 59

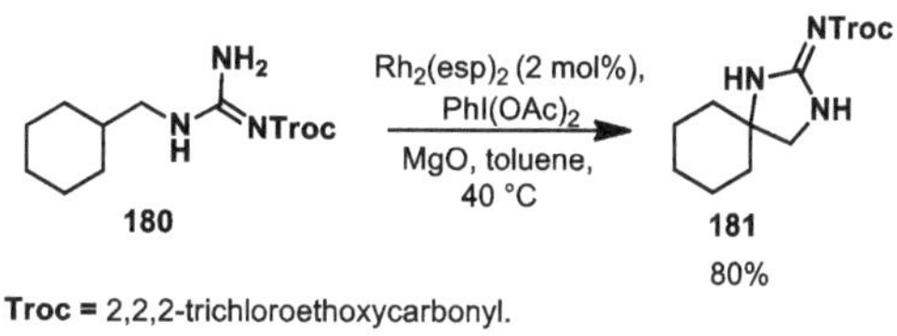


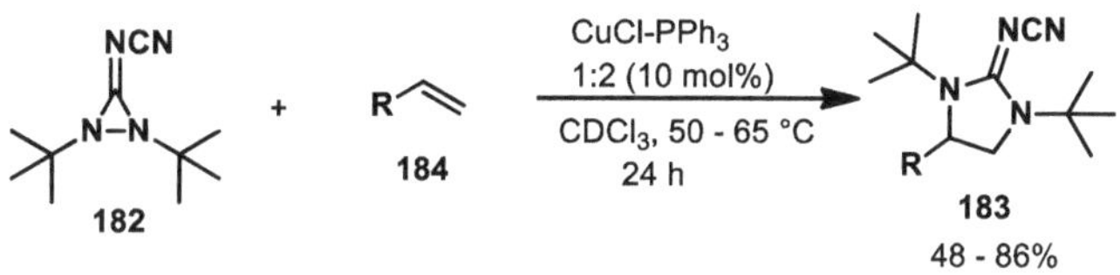


Scheme 60

## 3.2 Metal-Catalysed Ring Closure

In Du Bois' seminal work on stereospecific C–H insertion chemistry, a novel route for the synthesis of cyclic guanidines was designed [92]. Oxidative C–H amination of *N*-Troc (Troc = 2,2,2-trichloroethoxycarbonyl)-protected guanidines **180** to specifically generate five-membered ring derivatives **181** rather than six-membered rings was demonstrated (Scheme 59) [92].

This chemistry has a wide range of applications and has found use in both natural product synthesis [93] and medicinal chemistry [48]. The use of Troc as a protecting group along with $Rh_2(esp)_2$ in catalytic quantities proved essential for reactivity. This strategy was effectively employed in the synthesis of the guanidine natural product gonyautoxin [94].

The use of diaziridinimines **182** as precursors for guanidines is a facile and atom-economical route towards five-membered cyclic guanidines **183** (Scheme 60) [95]. Reacting **182** with a variety of olefins **184** in the presence of CuCl in catalytic quantities can afford the desired guanidine products, as described by Shi and

**Fig. 7** Proposed catalytic cycle of guanidylation [95]

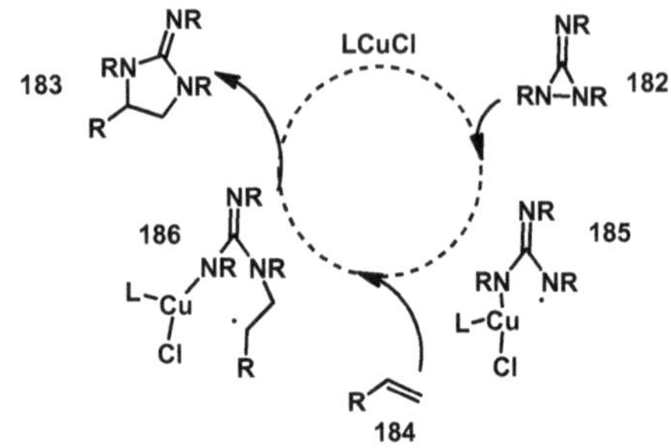


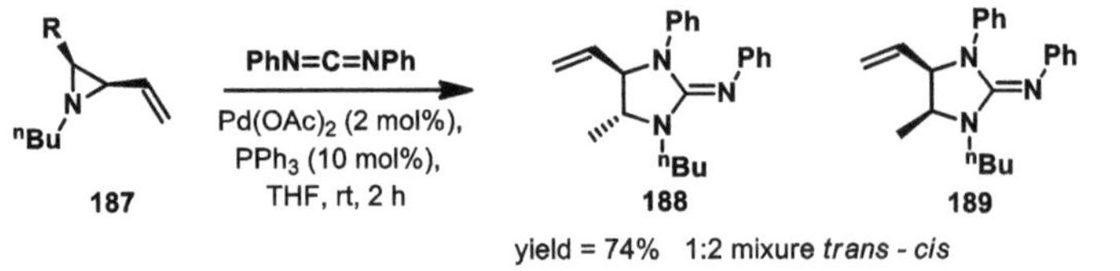


**Scheme 61**

**Scheme 62**

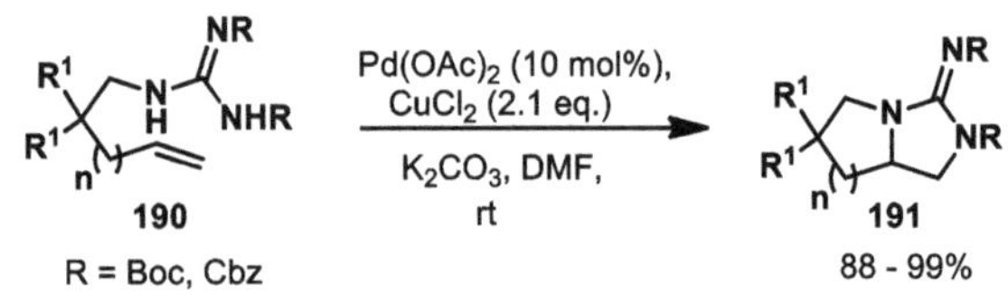


co-workers. Reactivity was shown to regioselectively proceed at terminal alkenes preferentially in the case of either dienes or trienes.

The reaction mechanism although still unknown is thought to proceed via homolytic cleavage of the N–N bond of **182** by CuCl, followed by addition of the nitrogen-centred radical **185** to olefin **184** and subsequent C–N bond formation **186** and regeneration of catalyst CuCl (Fig. 7).

The exposure of vinyl aziridines **187** to catalytic amounts of Pd and $PPh_3$ in the presence of carbodiimides led to the generation of cycloaddition products **188** and **189** (Scheme 61) [96]. Initial Pd-catalysed aziridine ring opening and subsequent $\pi$-allyl complex formation is presumably followed by guanidine formation and subsequent ring closure to yield the desired five-membered rings. The mixture of stereoisomers obtained suggests that there is a $\eta^3$–$\eta^1$–$\eta^3$ interconversion of a $\pi$-allyl Pd complex [97]. A recent report from the Stoltz laboratory presented a Lewis acid-mediated variation of this reaction, in which aziridines with no vinyl appendages are exposed to a variety of carbodiimides to produce iminoimidazolidines [98].

In 2008, Muñiz and co-workers described the direct synthesis of bicyclic guanidines through an unprecedented palladium(II)-catalysed diamination with copper chloride as reoxidant [99]. The synthesis of these bicycles occurs by means of a Pd-catalysed intramolecular guanidine transfer to alkenes **190** affording product **191** (Scheme 62). This method proves particularly effective in the formation of 5,5- and 5,6-bicyclic guanidines when guanidine is either Boc or Cbz protected.

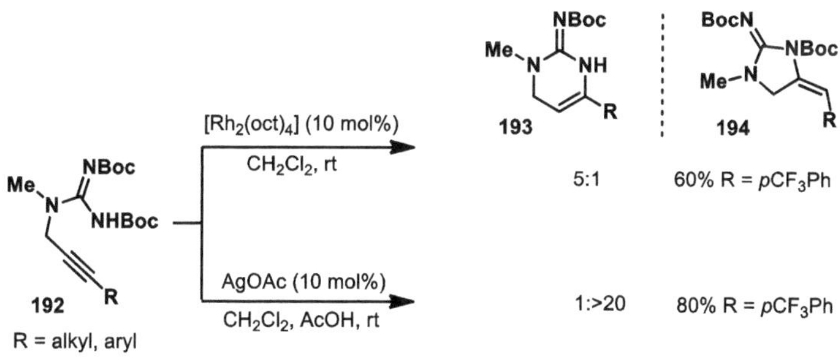


Scheme 63

Scheme 64

In 2011, a report by Looper and co-workers exploited the reactivity of propargylguanidines **192** in hydroamination reactions (Scheme 63) [100]. Two possible cyclisation pathways of propargylguanidine **192** can be promoted leading to both five- and six-membered cyclic guanidines (**193** and **194**, respectively). The reactivity of these starting materials can be adapted into either product by choice of catalyst.

Metal-catalysed cyclisations on alkynes typically favour 5-*exo*-dig pathways [101], and therefore, the ability to also form 6-*endo*-dig products is of particular interest. Looper et al. have made use of this methodology in their synthesis of saxitoxin [102]. During the preparation of the corresponding manuscript by Looper and co-workers, a similar type of reactivity was discovered by van der Eycken [103]. In this approach, propargylamines **195** were exposed to guanidylating conditions using $AgNO_3$ as the promoting agent (Scheme 64).

In the presence of silver(I) salts, a 5-*exo*-dig cyclisation was facilitated, and, in one pot, the transformation of propargylamines **195** into cyclic guanidines **196** was effected. Sterically hindered propargylamines as well as *N*-aryl propargylamines were also able to undergo this cyclisation pathway.

## 3.3 Six-Membered Cyclic Guanidines

Although not as widely researched as the formation of five-membered cyclic guanidine structures, the formation of six-membered rings has received attention in the synthesis of numerous natural products and biologically active compounds.

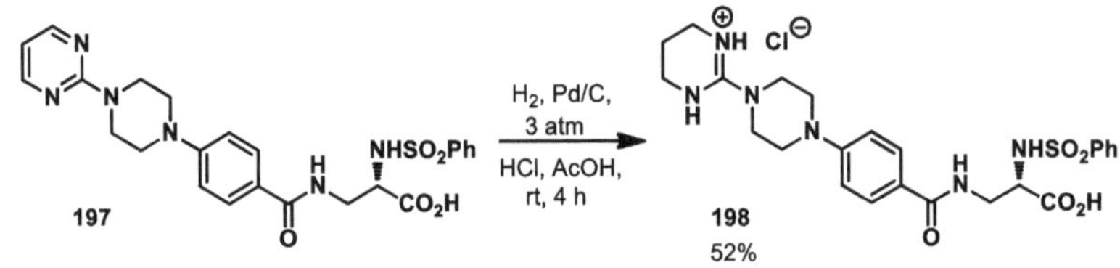

Scheme 65

Scheme 66

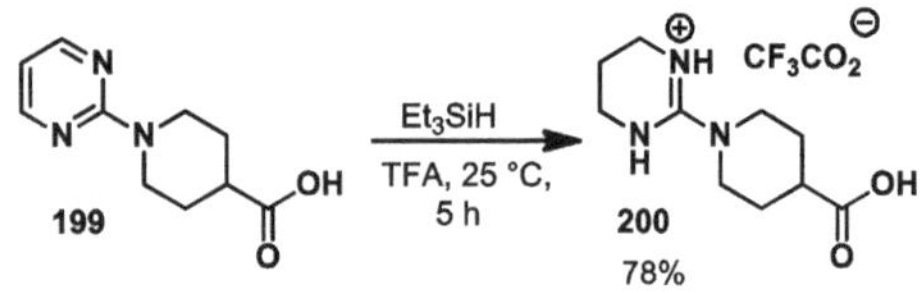

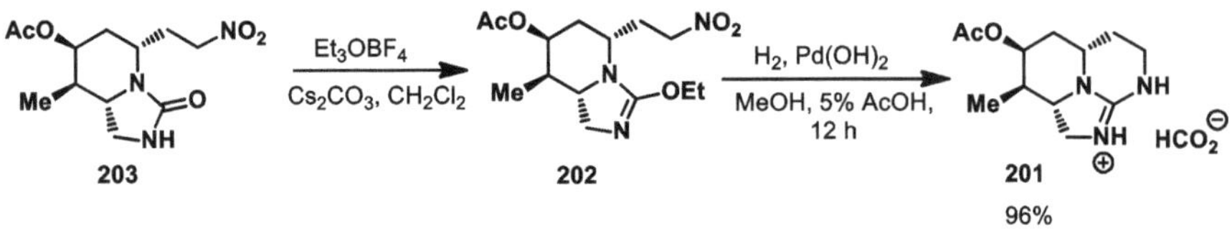

Scheme 67

The synthesis of *N*-substituted-2-amino-1,4,5,6-tetrahydropyrimidines (six-membered cyclic guanidines) has been reported by means of hydrogenation of the 2-aminopyrimidine precursor **197**. In a search for integrin $\alpha_v\beta_3$ antagonists, Ajito et al. employed hydrogenation under acidic conditions catalysed by Pd/C generating the cyclic guanidine **198** in an acceptable yield (Scheme 65) [104].

Another similar reductive method for the synthesis of *N*-substituted-2-amino-1,4,5,6-tetrahydropyrimidines from 2-aminopyrimidines was utilised by Shen and co-workers by using triethylsilane and TFA [105]. Initial experiments gave both dihydro- and tetrahydropyrimidine products; however, upon optimisation, tetrahydropyrimidines were generated in poor to excellent yields (25–90% yield). A representative example is shown in Scheme 66, with the conversion of amino pyrimidine **199** into guanidine **200**.

In an effort to synthesise tricyclic guanidine **201**, a reductive guanidylation method was formulated by Williams and co-workers [106]. Initial guanidine precursor **202** was synthesised from urea **203** using Meerwein's salt (triethyloxonium tetrafluoroborate). The alkyl nitro functionality was then reduced using palladium-catalysed hydrogenation to form the prerequisite primary amine which is now ideally located to displace ethoxide in an addition–elimination reaction (Scheme 67).

The facile synthesis of *N*-acyl cyanamides lends itself to an appealing method for the formation of tricylic guanidines **204** by means of radical domino cycloaddition (Scheme 68) [107]. This method, devised by Lacôte and co-workers, is the

Scheme 68

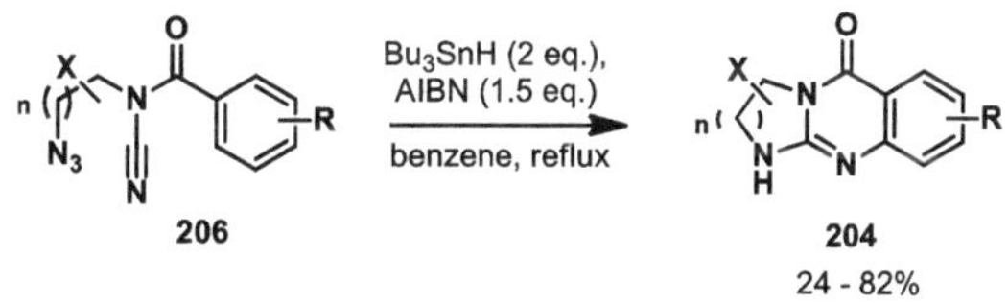


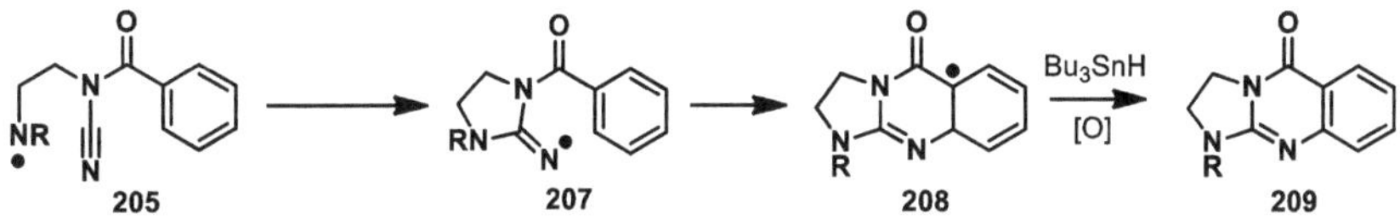


Scheme 69

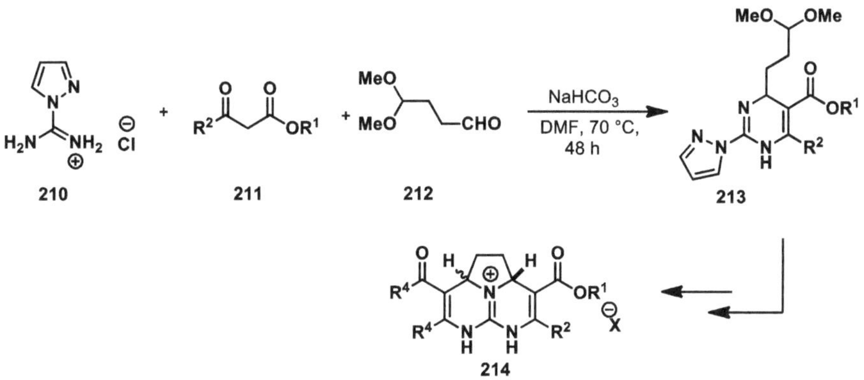


Scheme 70

sole literature example of the utilisation of radical chemistry to form cyclic guanidines.

Initial formation of a nitrogen-centred radical **205** from the prerequisite azide **206** and reaction with the adjacent cyanamide form a guanidine-type radical **207** (Scheme 69). Subsequent interaction with the π-system forms the final six-membered ring **208**, whilst the presence of $Bu_3SnH$ and an oxidant allows the re-aromatisation of the ring generating **209**.

Overman and co-workers have effectively utilised the three-component Biginelli reaction in their research towards the synthesis of six-membered cyclic guanidines [108]. Utilising pyrazole guanidine **210**, β-keto-ester **211** and derivatised aldehyde **212** in a three-component Biginelli reaction, the desired cyclic guanidines **213** were formed in good yield (Scheme 70). These products, after further functionalisation, were converted into tricycle **214**, a precursor for numerous guanidine natural products.

In an effort to synthesise similar tricyclic guanidine-containing compounds, Nagasawa et al. designed a guanidine condensation-type reaction leading to the

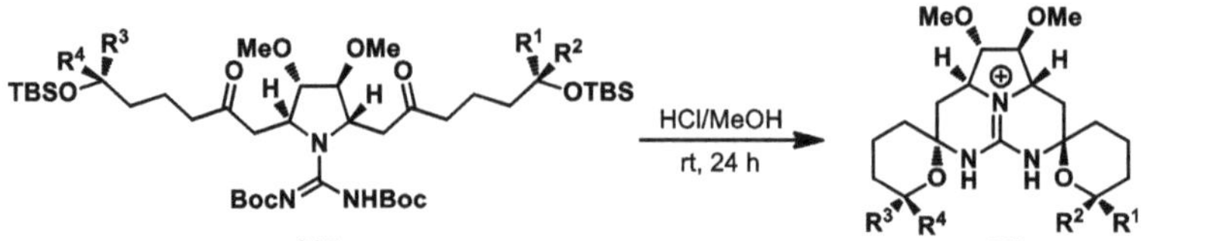


Scheme 71

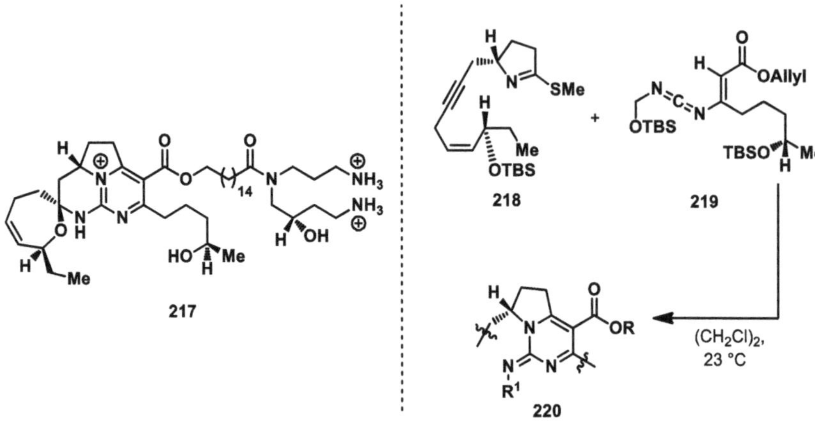


Scheme 72

generation of five rings in a single step [109]. This impressive reaction was facilitated by an acid-mediated deprotection of **215** and concomitant condensation of the guanidine and hydroxy functionalities onto the available ketone moieties (Scheme 71) to afford compound **216**.

In Gin's exquisite synthesis of crambidine (**217**) [110], the construction of the central cyclic guanidine was achieved by means of a [4 + 2] annulation between thioimidate **218** and vinyl carbodiimide **219** (Scheme 72). This route rapidly generated the core compound **220** which upon further functionalisation resulted in the formation of **217**. Another key step was the hydroamination of an alkyne moiety (present in substrate **218**) with derivatised 2-aminopyrimidine (present in substrate **220**) as the nucleophile.

In 2002, Isobe and co-workers described a synthesis of 11-deoxytetrodotoxin **221** incorporating an interesting method of cyclic guanidine formation (Scheme 73) [111]. The acetate-protected guanidine functionality present in precursor **222** had been installed using traditional methods. Deprotection of the acetate groups ($NH_4OH$, MeOH, $H_2O$) followed by exposure to acidic conditions (TFA, $H_2O$) resulted in orthoester formation and guanidine cyclisation, displacing a methoxy group and forming a hemiaminal. This is an impressive transformation due to the molecular complexity of both starting material and product.

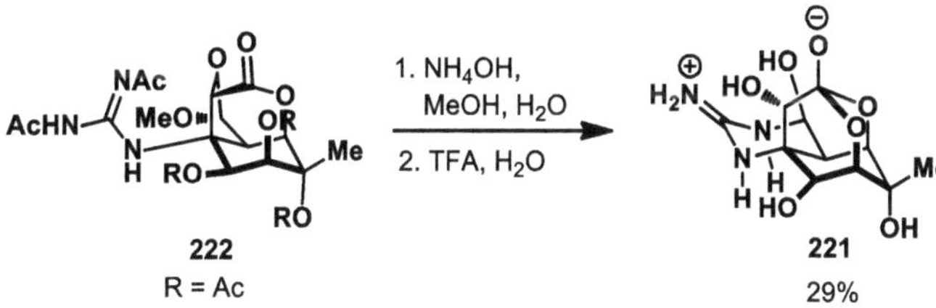


Scheme 73

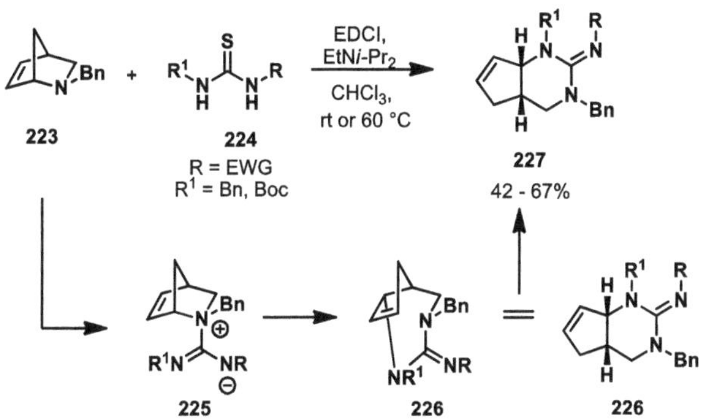


Scheme 74

It has been shown that bridged bicyclic allylic tertiary amines such as the aza-norbornene **223** can add to an in situ-generated carbodiimide (formed from thiourea derivative **224**) to form a zwitterionic intermediate **225** (Scheme 74) [112].

These strained systems are then ideally suited to undergo 1,3-diaza-Claisen rearrangement forming compound **226**. The presence of an electron-withdrawing substituent on the intermediate carbodiimide not only makes the species more reactive to nucleophilic attack but also stabilises the developing negative charge present in the zwitterionic species **225** (Scheme 74). A number of varying thioureas **224** were shown to be reactive towards carbodiimide formation when EDCI was used as a promoting agent. In Harran's study of the axinellamine natural products [113], a six-membered thiourea analogue was used to introduce the guanidine functionality. Initially, compound **227** was reacted with carboxylic acid **228** to form intermediate **229**, and, then, in the presence of oxalyl chloride, cyclisation occurred to afford tricylic compound **230** (Scheme 75). This presents an operationally simple and effective method for guanidine installation.

In the synthesis of the alkaloid alchorneine **231**, a palladium-mediated cyclisation of a prenyl functional group with cyclic guanidine **232** was developed (Scheme 76) [114]. Although 2 equiv. of palladium were required for the

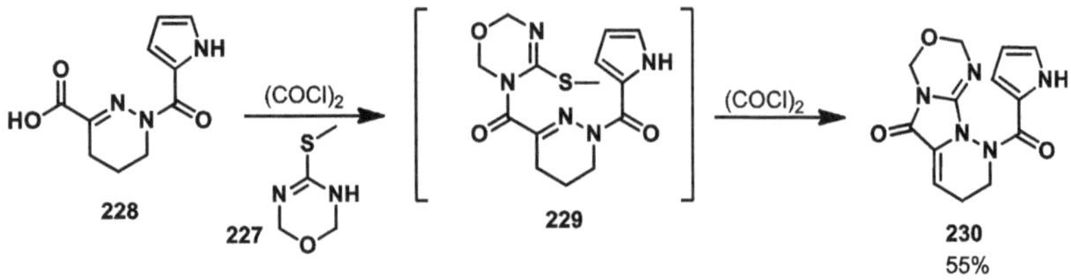


Scheme 75

Scheme 76

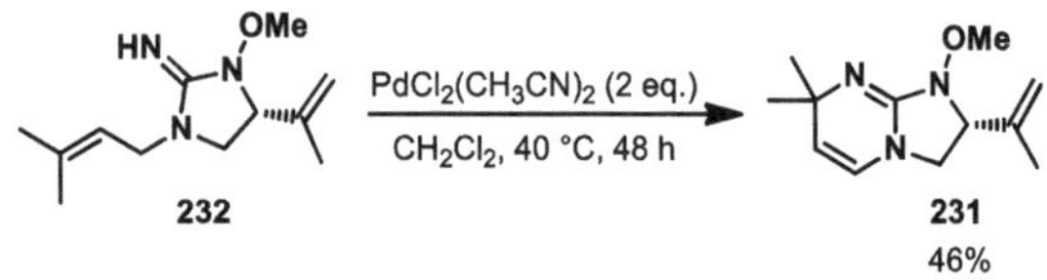


Scheme 77

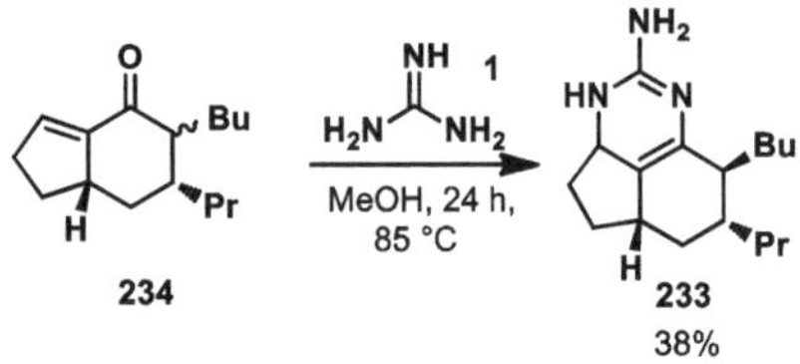


cyclisation to occur, this represents an interesting method for the formation of functionalised 1,4-dihydropyrimidines.

Snider and co-workers incorporated a Michael addition followed by a condensation reaction to form the tricyclic natural product netamine E **233** [115]. This direct approach facilitated the synthesis of a number of natural products containing the same core. When Michael acceptor **234** in methanol under reflux was combined with guanidine, compound **233** was produced in 38% yield (Scheme 77).

In an exceptionally elegant synthesis of saxitoxin, Bhonde and Looper designed a highly elaborate cascade reaction to generate the key intermediate **235** starting from acyclic guanidine **236** (Scheme 78) [102]. The cascade begins with Ag(I)-promoted ring closure of the benzyl-protected guanidine function onto the alkyne, generating a five-membered ring bearing an exocyclic olefin (Scheme 78, step 1). This olefin, under iodine activation (step 2), is then prone to nucleophilic attack by the acyclic guanidine functionality forming the six-membered ring (step 3). Ensuing expulsion of the resulting secondary alkyl iodide, facilitated by the use of Ag (I) and acetic acid, then generates highly functionalised intermediate **235**.

In another approach towards saxitoxin, Nishikawa and co-workers incorporated an ingenious cascade reaction in their synthetic route [116]. This allows the rapid generation of complexity in their system. Exposure of alkyne **237** to pyridinium hydrobromide perbromide ($PyHBr_3$) facilitates a bromocyclisation reaction that

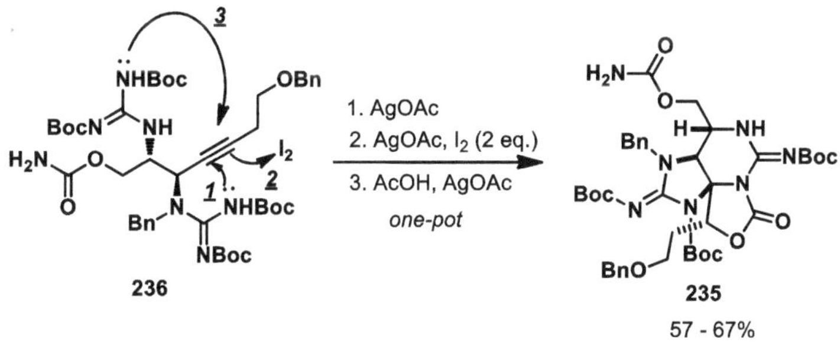


Scheme 78

Scheme 79

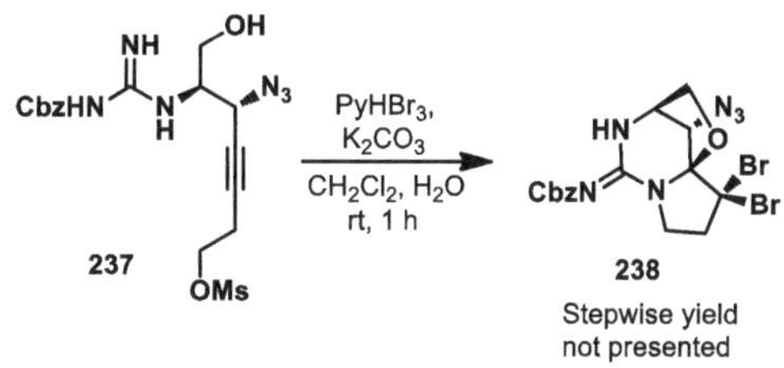


allows *N*-alkylation of the guanidine to occur (Scheme 79). This remarkable reaction generated the tricyclic core **238** in a highly stereospecific manner.

A recent method for the preparation of six-membered cyclic guanidines was devised by Rozas, Grayson and Shaw [117]. A series of aryl halides **239** were reacted with 2-aminopyrimidine **240** to afford *N*-substituted 2-aminopyrimidines **241** in good yields. These pyrimidine intermediates were then exposed to hydrogenation conditions to afford the reduced tetrahydropyrimidine analogues **242** (Scheme 80). This represents a facile two-step synthesis of *N*-substituted 2-amino-1,4,5,6-tetrahydropyrimidines **242**.

In a related publication by the same authors, aryl halides **239** were effectively coupled under Pd catalysis with 2-amino-4,6-dimethoxypyrimidine **243** to generate substituted pyrimidines **244** in moderate to excellent yields [118]. In an interesting ring cleavage reaction, **244** could be converted into guanidines **245** under acidic conditions (Scheme 81).

## 3.4 Seven-Membered Rings

In their synthesis of the bicyclic guanidine alkaloid (+)-monanchorin, Sutherland and co-workers reported a late-stage deprotection–cyclisation cascade to form the desired guanidine hemiaminal product **246** [119]. Acid-mediated deprotection of

Scheme 80

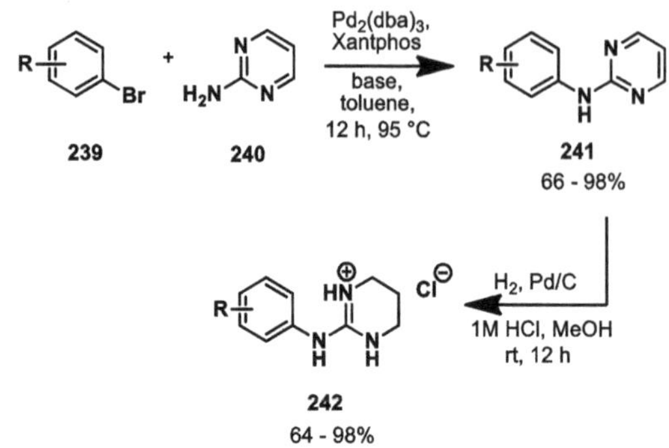


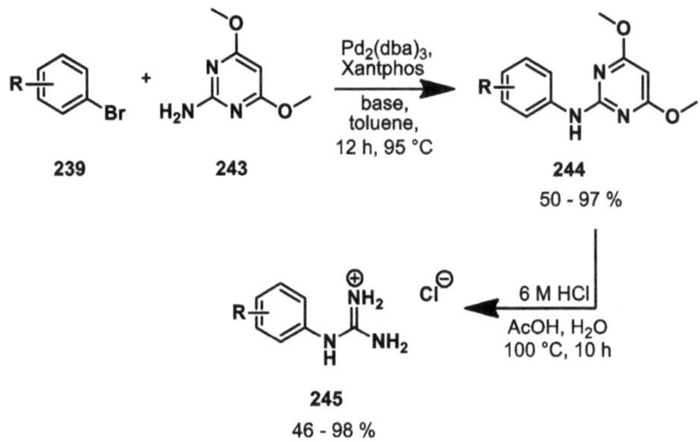


Scheme 81

Scheme 82

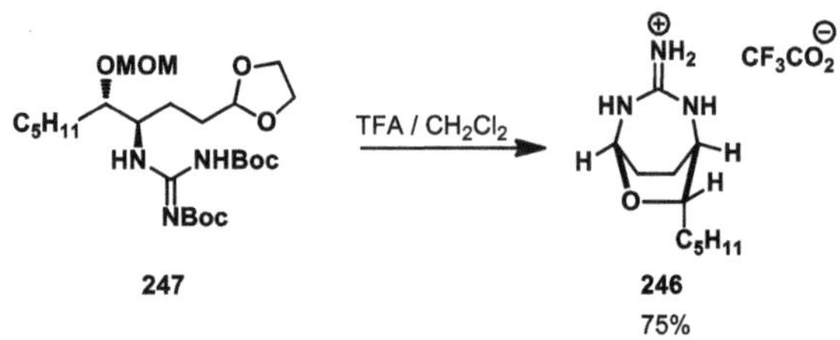


the aldehyde, hydroxy and guanidine functionalities of **247** in one pot allowed for an efficient cyclisation reaction pathway, forming an unusual seven-membered guanidine-containing ring (Scheme 82).

In an analogous reaction to the ring-opening/ring-closing reaction of aziridines with carbodiimides to form five-membered cyclic guanidines that has been

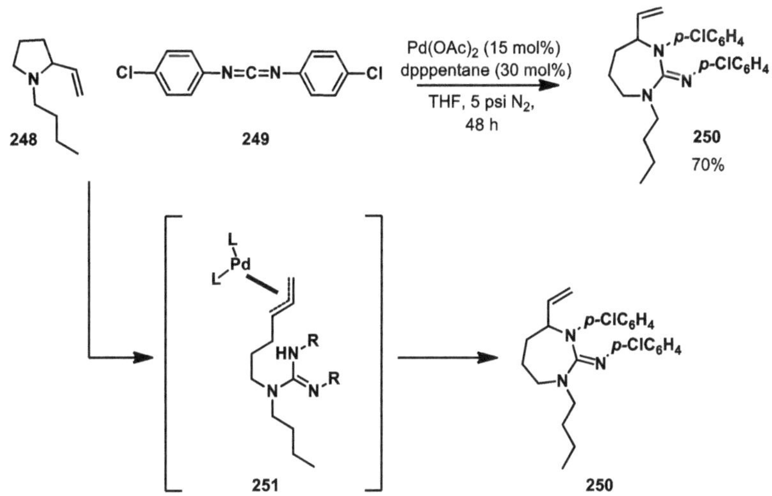


**Scheme 83**

discussed earlier, 2-vinylpyrrolidines **248** can react with carbodiimides **249** to form seven-membered cyclic guanidines **250** (Scheme 83) [120].

The reaction was shown to be optimal under slightly longer and harsher conditions (48 h, 5 psi $N_2$) than those required for the aziridine ring opening, presumably due to the inherent relief of ring strain associated with aziridine ring opening. Formation of the π-allyl Pd complex **251** followed by guanidine-induced ring closure and β-hydride elimination generated the desired vinyl guanidines **250**.

# 4 Some Biological Applications of Guanidines

We have so far shown different possibilities for the synthesis of guanidine derivatives, and this is highly relevant because, despite the high basicity of this function that could challenge bioavailability of a drug, guanidine groups are present in many important therapeutic agents already in the market such as metformin (type II diabetes), zanamivir (anti-influenza), guanfacine (attention deficit hyperactivity disorder), cimetidine and famotidine (both antacid), to mention just a few (Fig. 8).

Several reports have appeared in the literature dealing with the biological activities of guanidine-containing compounds, such as those published by Sączewski and Balewski in 2009 and 2013 dealing with patents and articles issued before 2008 [121] and the patents filed during the period 2008–2012 [13]. To attempt an in-depth review of this topic is out of the scope of the present chapter; however, we will present a short overview of the work performed in our laboratory in relation to biologically active guanidine-like derivatives.

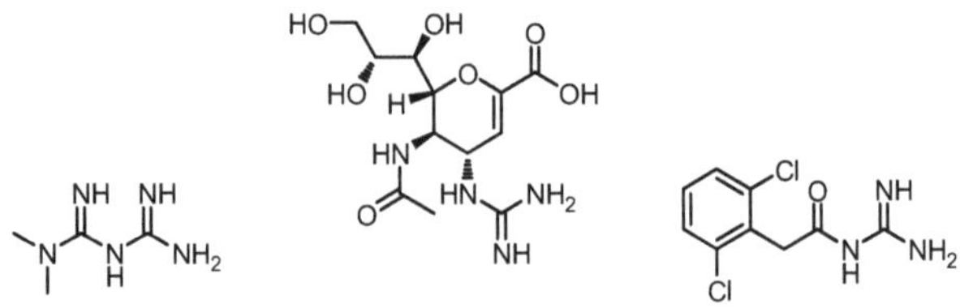

252 Metformin

253 Zanamivir

254 Guanfacine

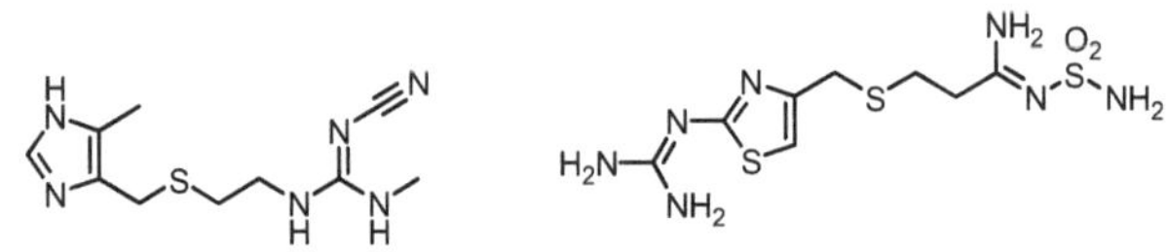

255 Cimetidine

256 Famotidine

**Fig. 8** Structure of guanidine-containing drugs currently on the market

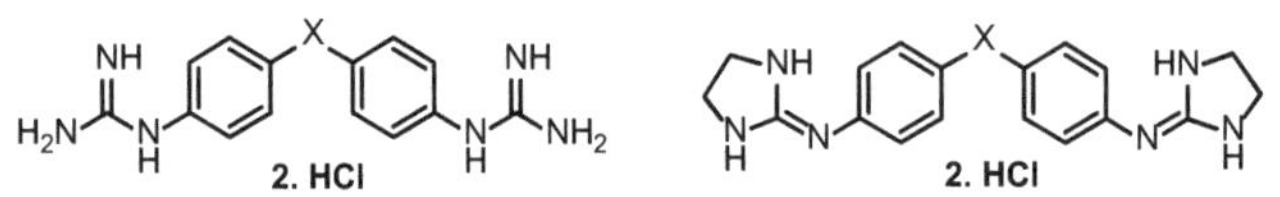


257

258

X = O (a), CO (b), $CH_2$ (c), NH (d), $CH_2CH_2$ (e), S (f), NHCO (g), etc

**Fig. 9** First generation of DNA minor groove binders prepared by Rozas and co-workers

## *4.1 Guanidine Derivatives as DNA Minor Groove Binders and Kinase Inhibitors*

During the last 15 years, we have focused on the design, synthesis and biological evaluation of novel DNA minor groove binders (MGBs). We found that several diphenyl 4,4′-bis-guanidinium (**257**) and -bis-2-iminoimidazolinidium (**258**) derivatives (Fig. 9) showed correlation between their DNA-binding affinity and their in vitro and in vivo antitrypanosomal activity, suggesting that DNA minor groove binding could be a mechanism of action of these compounds [122].

Based on these results showing DNA targeting, new derivatives of 4,4′-asymmetrical diaromatic guanidinium/2-iminoimidazolinidium (**259**) [123] and guanidinium/aminoalkyl carbamide (**260**) [124] as well as 4,4′-symmetrical bis-2-amino-(1,4,5,6-tetrahydropyrimidinium) (**261**) [125], bis-isouronium (**262**) [126] and bis-hydroxyguanidinium (**263**) [127] derivatives were synthesised (Fig. 10). The affinity of these substrates for DNA was evaluated by means of DNA thermal denaturation experiments and other biophysical methods (circular and linear dichroism, isothermal calorimetry, UV titration methods, surface plasmon

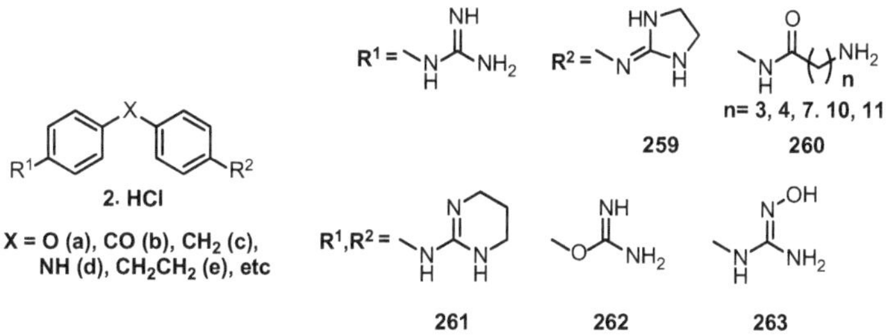


**Fig. 10** Different guanidine-like families of compounds prepared as DNA minor groove binders

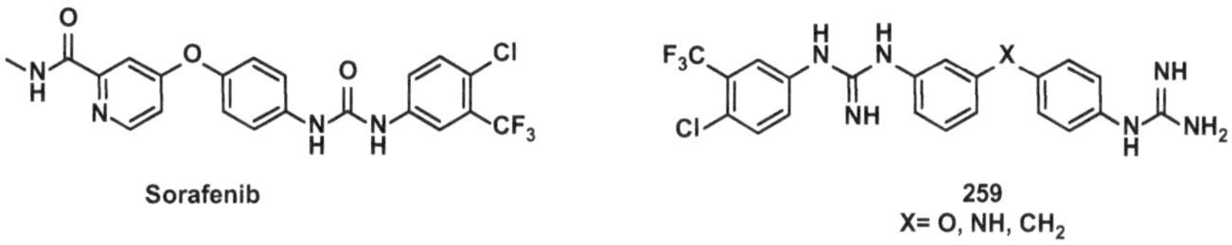


**Fig. 11** Sorafenib and some of Rozas' guanidine derivatives showing activity as BRAF inhibitors

resonance), finding that most of these compounds strongly bind to the minor groove of DNA [127–129].

These families produced encouraging results not only in terms of binding affinity and selectivity towards the minor groove of DNA but also in terms of antiparasitic [124] and anticancer activity as well as induction of cellular apoptosis [123, 127, 130]. Thus, very good cytotoxicity was found for these 4,4′-bis-guanidinium-like derivatives in human leukaemia (HL-60), breast cancer (MCF-7) and neuroblastoma (Kelly) cell lines.

However, some contradictions were found between the biophysical and the biological activity of these families of compounds because the most cytotoxic and apoptosis-inducing compounds (**257a** and **258a**) were the weakest DNA minor groove binders and also some of the least biologically potent compounds bind the strongest to DNA (**257b** and **258b**) [123, 126]. This suggests that the mechanism of action in inducing apoptosis of the former could take place by a different mechanism than DNA targeting. Hence, looking for alternative mechanisms of action, we noticed several structural similarities between these compounds and some protein kinase inhibitors (e.g. sorafenib, Fig. 11), and accordingly, we prepared new guanidine derivatives more related to these kinase inhibitors (see compounds **10** in Scheme 4 and compounds **259** in Fig. 11). Some of the new guanidine derivatives prepared have shown very good inhibition of the BRAF kinase (>98%) as well as cytotoxic activity of HL-60, MCF-7 and colorectal cancer (RKO) cell lines [19, 131].

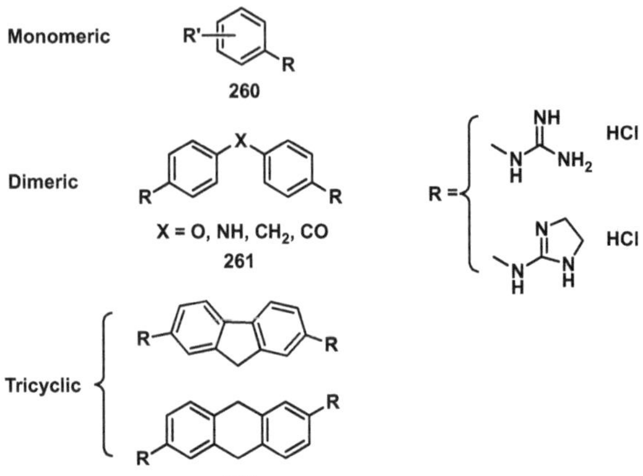


**Fig. 12** Different cores used for the guanidinium and 2-aminoimidazolinium derivatives prepared by Rozas and co-workers

## *4.2 Guanidines Targeting $\alpha_2$-Noradrenaline Receptors*

In the search of novel neuropsychiatric therapies, research within our group has described to date the synthesis, pharmacological evaluation and structure–activity relationships of over 100 molecules targeting the $\alpha_2$-noradrenaline receptors (α2-AR) [77, 132, 133]. The α2-AR are involved in a number of neurological disorders such as depression or schizophrenia. All antidepressants currently in the market act by increasing the levels of monoamines such as serotonin or noradrenaline (NA), what has recently been demonstrated to improve neuroplasticity (changes in neuronal pathways and synapses) [134]. We are aiming to prepare α2-AR antagonist with the idea of increasing the levels of NA in the brain resulting in improvements in neuroplasticity.

The first molecules prepared had the general structure of a substituted phenyl ring directly attached to either a guanidine or 2-iminoimidazolidine group and were prepared as the corresponding hydrochlorides due to the water solubility of these salts which is desirable for pharmacological testing (**260**, Fig. 12). Several related dimeric molecules have also been investigated, where the phenyl rings are attached via a linker (X) and the guanidine-like cations are connected in a 4,4′-disposition (**261**, Fig. 12). Some tricyclic derivatives were also synthesised (**262**, Fig. 12).

All these molecules were evaluated in vitro for their α2-AR affinity using a competition binding assay in human prefrontal cortex (PFC) of depressed suicide victims; the most successful derivatives from these tests ($pK_i > 6.5$) were subjected to both in vitro ([$^{35}$S]GTPγS assays in human PFC) and in vivo (microdialysis in rats) to determine their activity as agonists or antagonist. Microdialysis studies measure NA concentrations in a conscious mouse and let us learn that both the direct administration of our compounds to the PFC and administration in the periphery led to increased extracellular concentrations of NA, meaning that these

**Fig. 13** α2-AR antagonists with antidepressant characteristics developed by Rozas and co-workers

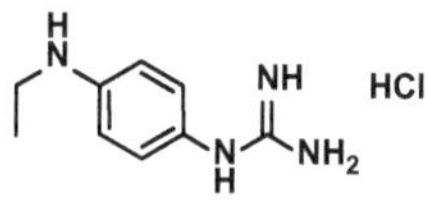


263

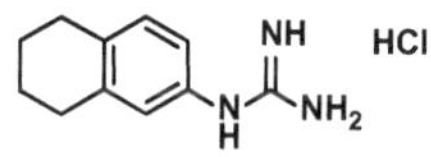


264

compounds can pass the blood–brain barrier. Several extremely high-affinity compounds were also obtained with $pK_i$ values close to 9.0 indicating near nanomolar binding affinity; however, these compounds had the undesired agonistic activity [77, 132, 133]. Finally, the best antagonists were tested in behavioural experiments in mice (tail suspension test (TST) and force swimming test (FST)) to determine their antidepressant effects. Only compounds **263** and **264** (Fig. 13) showed antidepressant effects in the TST and FST outperforming fluoxetine and mirtazapine (known commercial antidepressants) [135].

However, rational structure–activity relationships (SARs) were difficult to obtain since minor structural changes had huge effects on binding affinity and activity. Overall, 2-iminoimidazolidine derivatives show higher affinity for the α2-AR than their guanidine analogues; however, no 2-iminoimidazolidines gave antidepressant effects in the behavioural tests, and the α2-AR activity of closely related analogues is unpredictable. For that reason, we approach a more rational design for these compounds, and we found that a second *N*-substitution in the guanidine moiety of aryl guanidines as well as the introduction of a pyridine core instead of a phenyl ring (see compounds **11** in Scheme 4) always results in compounds with α2-AR antagonistic activity even though the corresponding affinity decreases in many cases [20, 136]. We are still exploring structural changes in these compounds that will improve both α2-AR affinity and antagonistic activity to achieve an optimal antidepressant with less secondary effects [137].

# 5 Conclusions

From the vast array of methodologies available for guanidine synthesis, it is evident that guanidine chemistry is a flourishing area of chemistry with a very promising future in the pharmaceutical industry. The presence of guanidines in this economically important area ensures the constant need for ever more specific and effective methods of guanidylation. But still there is need for improvements in the field. For instance, the direct coupling of unprotected guanidines to substrates would be a highly useful methodology allowing for late-stage guanidine installation. No doubt that in the coming years, problems such as this and many more in guanidine synthesis will be answered.

## References


1. Menor-Salvan C, Marin-Yaseli MR (2012) Chem Soc Rev 41:5404–5415
2. Haas DJ, Harris DR, Mills HH (1965) Acta Crystallogr 19:676–679
3. Yamada T, Liu X, Englert U, Yamane H, Dronskowski R (2009) Chem Eur J 15:5651–5655
4. Parker EJ, Pratt AJ (2010) In: Hughes AB (ed) Amino acids, peptides, proteins in organic chemistry. Wiley-VCH, Weinheim, pp 1–82
5. Sokalingam S, Raghunathan G, Soundrarajan N, Lee SG (2012) PLoS One 7:e40410
6. Dougherty DA (2013) Acc Chem Res 46:885–893
7. Crowley PB, Golovin A (2005) Proteins 59:231–239
8. Blanco F, Kelly B, Sanchez-Sanz G, Trujillo C, Alkorta I, Elguero J, Rozas I (2013) J Phys Chem B 11:11608–11616
9. Kelly B, Sanchez-Sanz G, Blanco F, Rozas I (2012) Comput Theor Chem 998:64–73
10. Gund P (1972) J Chem Ed 49:100–103
11. Blanco F, Kelly B, Alkorta I, Rozas I, Elguero J (2011) Chem Phys Lett 511:129–134
12. Rozas I, Sanchez-Sanz G, Alkorta I, Elguero J (2013) J Phys Org Chem 26:378–385
13. Sączewski F, Balewski Ł (2013) Expert Opin Ther Pat 23:965–995
14. Berlinck RG, Trindade-Silva AE, Santos MF (2012) Nat Prod Rep 29:1382–1406
15. Kishi Y, Aratani M, Fukuyama T, Nakatsubo F, Goto T, Inoue S, Tanino H, Sugiura S, Kakoi H (1972) J Am Chem Soc 94:9217–9219
16. Suhs T, Konig B (2006) Mini Rev Org Chem 3:315–331
17. Kim KS, Qian LG (1993) Tetrahedron Lett 34:7677–7680
18. Chen H-M, Li G, Cao L-H (2008) J Chin Chem Soc 55:474–478
19. Diez-Cecilia E, Kelly B, Perez C, Zisterer DM, Nevin DK, Lloyd DG, Rozas I (2014) Eur J Med Chem 81:427–441
20. Kelly B, McMullan M, Muguruza C, Ortega JE, Meana JJ, Callado LF, Rozas I (2015) J Med Chem 58:963–977
21. Cunha S, Rodrigues MT, da Silva CC, Napolitano HB, Vencato I, Lariucci C (2005) Tetrahedron 61:10536–10540
22. O'Donovan DH, Rozas I (2011) Tetrahedron Lett 52:4117–4119
23. Cunha S, Rodrigues MT (2006) Tetrahedron Lett 47:6955–6956
24. Kelly B, Rozas I (2013) Tetrahedron Lett 54:3982–3984
25. Ramadas K, Srinivasan N (1995) Tetrahedron Lett 36:2841–2844
26. Shibanuma T, Shiono M, Mukaiyama T (1977) Chem Lett 575–576
27. Ohara K, Vasseur JJ, Smietana M (2009) Tetrahedron Lett 50:1463–1465
28. Porcheddu A, De Luca L, Giacomelli G (2009) Synlett 3368–3372
29. Maryanoff CA, Stanzione RC, Plampin JN, Mills JE (1986) J Org Chem 51:1882–1884
30. Isidro-Llobet A, Alvarez M, Albericio F (2009) Chem Rev 109:2455–2504
31. Manimala JC, Anslyn EV (2002) Tetrahedron Lett 43:565–567
32. Madalengoitia J, Flemer S (2007) Synthesis 1848–1860
33. Alonso-Moreno C, Antinolo A, Carrillo-Hermosilla F, Otero A (2014) Chem Soc Rev 43: 3406–3425
34. Kantam ML, Priyadarshini S, Joseph PJA, Srinivas P, Vinu A, Klabunde KJ, Nishina Y (2012) Tetrahedron 68:5730–5737
35. Pottabathula S, Royo B (2012) Tetrahedron Lett 53:5156–5158
36. Toy PH, Lam Y (eds) (2012) Solid-phase organic synthesis: concepts, strategies, applications. Wiley, Hoboken
37. Drewry DH, Gerritz SW, Linn JA (1997) Tetrahedron Lett 38:3377–3380
38. Josey JA, Tarlton CA, Payne CE (1998) Tetrahedron Lett 39:5899–5902
39. Ube H, Uraguchi D, Terada M (2007) J Organomet Chem 692:545–549
40. Powell DA, Ramsden PD, Batey RA (2003) J Org Chem 68:2300–2309
41. Ma D, Xia C, Jiang J, Zhang J, Tang W (2003) J Org Chem 68:442–451
42. Kim M, Mulcahy JV, Espino CG, Bois JD (2006) Org Lett 8:1073–1076

43. Roizen JL, Zalatan DN, Bois JD (2013) Angew Chem Int Ed Engl 52:11343–11346
44. Wang S, Romo D (2008) Angew Chem Int Ed Engl 47:1284–1286
45. Feichtinger K, Zapf C, Sings HL, Goodman M (1998) J Org Chem 63:3804–3805
46. Baker TJ, Rew Y, Goodman M (2000) Pure Appl Chem 72:347–354
47. Wu YQ, Hamilton SK, Wilkinson DE, Hamilton GS (2002) J Org Chem 67:7553–7556
48. Bernatowicz MS, Wu YL, Matsueda GR (1992) J Org Chem 57:2497–2502
49. Yong YF, Kowalski JA, Thoen JC, Lipton MA (1999) Tetrahedron Lett 40:53–56
50. Musiol HJ, Moroder L (2001) Org Lett 3:3859–3861
51. Gagnon PE, Boivin JL, Dickson JH (1959) Can J Chem 37:520–524
52. Rachlin S, Bramm E, Ahnfelt-Ronne I, Arrigoni-Martelli E (1980) J Med Chem 23:13–20
53. Wang Z (2010) Comprehensive organic name reactions, reagents. Wiley, Hoboken
54. Elliott AJ, Morris PE Jr, Petty SL, Williams CH (1997) J Org Chem 62:8071–8075
55. Atwal KS, Ferrara FN, Ahmed SZ (1994) Tetrahedron Lett 35:8085–8088
56. Mitsunobu O, Yamada Y (1967) Bull Chem Soc (Japan) 40:2380–2382
57. Dembinski R (2004) Eur J Org Chem 2004:2763–2772
58. Feichtinger K, Sings HL, Baker TJ, Matthews K, Goodman M (1998) J Org Chem 63: 8432–8439
59. Fishlock D, Guillemette JG, Lajoie GA (2002) J Org Chem 67:2352–2354
60. Olivier KS, Van Nieuwenhze MS (2010) Org Lett 12:1680–1683
61. Jacobson GB, Westerberg G, Markides KE, Långström B (1996) J Am Chem Soc 118: 6868–6872
62. Ha HH, Kim JS, Kim BM (2008) Bioorg Med Chem Lett 18:653–656
63. Looper RE, Haussener TJ, Mack JB (2011) J Org Chem 76:6967–6971
64. Beletskaya IP, Cheprakov AV (2004) Coord Chem Rev 248:2337–2364
65. Sambiagio C, Marsden SP, Blacker AJ, McGowan PC (2014) Chem Soc Rev 43:3525–3550
66. Zhang H, Cai Q, Ma D (2005) J Org Chem 70:5164–5173
67. Cohen T, Wood J, Dietz AG (1974) Tetrahedron Lett 15:3555–3558
68. Paine AJ (1987) J Am Chem Soc 109:1496–1502
69. Deng X, McAllister H, Mani NS (2009) J Org Chem 74:5742–5745
70. Cortes-Salva M, Nguyen BL, Cuevas J, Pennypacker KR, Antilla JC (2010) Org Lett 12: 1316–1319
71. Hammoud H, Schmitt M, Bihel F, Antheaume C, Bourguignon JJ (2012) J Org Chem 77: 417–423
72. Beletskaya IP, Cheprakov AV (2012) Organometallics 31:7753–7808
73. Xing H, Zhang Y, Lai Y, Jiang Y, Ma D (2012) J Org Chem 77:5449–5453
74. Li J, Neuville L (2013) Org Lett 15:6124–6127
75. Miyabe H, Yoshida K, Reddy VK, Takemoto Y (2009) J Org Chem 74:305–311
76. Dardonville C, Goya P, Rozas I, Alsasua A, Martin I, Borrego J (2000) Bioorg Med Chem 8:1567–1577
77. Rodriguez F, Rozas I, Ortega JE, Meana JJ, Callado LF (2007) J Med Chem 50:4516–4527
78. Kan WM, Lin SH, Chern CY (2005) Synth Commun 35:2633–2639
79. Hensler ME, Bernstein G, Nizet V, Nefzi A (2006) Bioorg Med Chem Lett 16:5073–5079
80. McKay AF, Kreling ME (1957) Can J Chem 35:1438–1445
81. European Patent, EP01986801990
82. Ye WP, Leow DS, Goh SLM, Tan CT, Chian CH, Tan CH (2006) Tetrahedron Lett 47: 1007–1010
83. Corey EJ, Grogan MJ (1999) Org Lett 1:157–160
84. Yamamoto Y, Mizuno H, Tsuritani T, Mase T (2009) Tetrahedron Lett 50:5813–5815
85. Li J (2009) Name reactions. Springer, Berlin, pp 332–333
86. González-Rosende ME, Castillo E, Asíns B, Mamouni R, Sepúlveda-Arques J (2007) Tetrahedron 63:8709–8714
87. Schroif-Gregoire C, Travert N, Zaparucha A, Al-Mourabit A (2006) Org Lett 8:2961–2964
88. Zhou L, Chen J, Zhou J, Yeung YY (2011) Org Lett 13:5804–5807

89. Bera S, Wallimann T, Ray S, Ray M (2008) FEBS J 275:5899–5909
90. Guiheneuf S, Paquin L, Carreaux F, Durieu E, Meijer L, Bazureau JP (2012) Org Biomol Chem 10:978–987
91. Li CM, Danishefsky SJ (2006) Tetrahedron Lett 47:385–387
92. Olson DE, Roberts DA, Du Bois J (2012) Org Lett 14:6174–6177
93. Hinman A, Du Bois J (2003) J Am Chem Soc 125:11510–11511
94. Mulcahy JV, Du Bois J (2008) J Am Chem Soc 130:12630–12631
95. Zhao B, Du H, Shi Y (2008) Org Lett 10:1087–1090
96. Butler DCD, Inman GA, Alper H (2000) J Org Chem 65:5887–5890
97. Godleski SA (1991) In: Trost BM, Fleming I, Semmelhack MF (eds) Comprehensive organic synthesis, vol 4. Pergamon, Oxford, pp 585–662, Chapter 3.3
98. Craig RA 2nd, O'Connor NR, Goldberg AF, Stoltz BM (2014) Chem Eur J 20:4806–4813
99. Hovelmann CH, Streuff J, Brelot L, Muniz K (2008) Chem Commun 2334–2336
100. Gainer MJ, Bennett NR, Takahashi Y, Looper RE (2011) Angew Chem Int Ed Engl 50: 684–687
101. Ritter S, Horino Y, Lex J, Schmalz HG (2006) Synlett 3309–3313
102. Bhonde VR, Looper RE (2011) J Am Chem Soc 133:20172–20174
103. Pereshivko OP, Peshkov VA, Ermolatev DS, van Hove S, Van Hecke K, Van Meervelt L, van der Eycken EV (2011) Synthesis 1587–1594
104. Ishikawa M, Tsushima M, Kubota D, Yanagisawa Y, Hiraiwa Y, Kojima Y, Ajito K, Anzai N (2008) Org Proc Res Dev 12:596–602
105. Baskaran S, Hanan E, Byun D, Shen W (2004) Tetrahedron Lett 45:2107–2111
106. Looper RE, Runnegar MTC, Williams RM (2006) Tetrahedron 62:4549–4562
107. Larraufie MH, Ollivier C, Fensterbank L, Malacria M, Lacote E (2010) Angew Chem Int Ed Engl 49:2178–2181
108. Nilsson BL, Overman LE (2006) J Org Chem 71:7706–7714
109. Nagasawa K, Georgieva A, Takahashi H, Nakata T (2001) Tetrahedron 57:8959–8964
110. Perl NR, Ide ND, Prajapati S, Perfect HH, Duron SG, Gin DY (2010) J Am Chem Soc 132: 1802–1803
111. Nishikawa T, Asai M, Isobe M (2002) J Am Chem Soc 124:7847–7852
112. Aranha Potter R, Bowser AM, Yang Y, Madalengoitia JS, Ziller JW (2013) J Org Chem 78: 11772–11782
113. Ding H, Roberts AG, Harran PG (2012) Angew Chem Int Ed Engl 51:4340–4343
114. Buchi G, Rodriguez AD, Yakushijin K (1989) J Org Chem 54:4494–4496
115. Yu M, Pochapsky SS, Snider BB (2008) J Org Chem 73:9065–9074
116. Sawayama Y, Nishikawa T (2011) Angew Chem Int Ed Engl 50:7176–7178
117. Shaw JW, Grayson DH, Rozas I (2014) Eur J Org Chem 161–174
118. Shaw JW, Grayson DH, Rozas I (2014) Arkivoc 3565–3569
119. Zaed AM, Sutherland A (2010) Org Biomol Chem 8:4394–4399
120. Zhou HB, Alper H (2004) Tetrahedron 60:73–79
121. Sączewski F, Balewski Ł (2009) Expert Opin Ther Pat 19:1417–1448
122. Dardonville C, Barrett MP, Brun R, Kaiser M, Tanious F, Wilson WD (2006) J Med Chem 49:3748–3752
123. Nagle PS, Rodriguez F, Kahvedzic A, Quinn SJ, Rozas I (2009) J Med Chem 52:7113–7121
124. McKeever C, Kaiser M, Rozas I (2013) J Med Chem 56:700–711
125. O'Sullivan P, Rozas I (2014) ChemMedChem 9:2063–2073
126. Goonan Á, Kahvedzic A, Rodriguez F, Nagle PS, McCabe T, Rozas I, Erdozain AM, Meana JJ, Callado LF (2008) Bioorg Med Chem 16:8210–8217
127. Kahvedzic A, Nathwani S-M, Zisterer D, Rozas I (2013) J Med Chem 56:451–459
128. Nagle PS, Rodriguez F, Quinn SJ, O'Donovan DH, Kelly JM, Nguyen B, Wilson WD, Rozas I (2010) Org Biomol Chem 8:5558–5567
129. Nagle PS, Rodriguez F, Nguyen B, Wilson WD, Rozas I (2012) J Med Chem 55:4397–4406
130. Kahvedzic A, Nathwani S-M, Zisterer D, Rozas I (2015)

131. Diez-Cecilia E, Carson R, Kelly B, van Schaybroeck S, Murray JT, Rozas I (2015)
132. Rodriguez F, Rozas I, Ortega JE, Erdozain AM, Meana JJ, Callado LF (2008) J Med Chem 51:3304–3312
133. Rodriguez F, Rozas I, Erdozain AM, Meana JJ, Callado LF (2009) J Med Chem 52:601–609
134. Nakamura S (2012) Antidepressants and morphological plasticity of monoamine neurons. In: Lu R-B (ed) Effects of antidepressants. InTech, Rijeka. ISBN 978-953-51-0663-0
135. Muguruza C, Rodriguez F, Rozas I, Meana JJ, Uriguen L, Callado LF (2013) Neuropharmacology 65:13–19
136. O'Donovan DH, Muguruza C, Callado LF, Rozas I (2014) Eur J Med Chem 82:242–254
137. McMullan M, Kelly B, Erdozain AM, Callado LF, Rozas I (2015)
138. Nagle PS, McKeever C, Rodriguez F, Nguyen B, Wilson WD, Rozas I (2014) J Med Chem 57:4397–4406

Top Heterocycl Chem (2017) 50: 53–70
DOI: 10.1007/7081_2015_169

Published online: 4 September 2015

# Bifunctional Guanidine Hydroxide and Related Organocatalysts

**Tsutomu Ishikawa**

**Abstract** Catalysts designed to connect a guanidine moiety and active hydrogens such as hydroxyl, amide, and amine functionalities through a chiral spacer will be discussed regarding their utilities in asymmetric synthesis. The active hydrogens coordinate with acceptor molecules through hydrogen bonding, thus achieving both activation and positional control in the transition states.

**Keywords** Asymmetric synthesis · Bifunctional guanidine · Hydrogen bonding · Organocatalysts

## Contents



T. Ishikawa (✉)
Graduate School of Pharmaceutical Sciences, Chiba University, 1-8-1 Inohana, Chuo, Chiba 260-8675, Japan
e-mail: benti@faculty.chiba-u.jp

# 1 Introduction

Chiral guanidines bearing an active hydrogen atom at the appropriate position would be expected to coordinate with acceptor groups through hydrogen bonding, thus achieving both activation and positional control. Enolate anions generated by deprotonation by the guanidine Brønsted bases would be controlled in their position and direction through hydrogen bonds with the two acidic hydrogen atoms of the resulting guanidinium cation. In this chapter, catalysts designed to connect a guanidine with active hydrogens such as hydroxyl, amide, and amine functionalities through a chiral spacer will be discussed regarding their utilities in asymmetric synthesis.

# 2 Guanidine Hydroxide Catalysts

## *2.1 Chiba-G and Related 2-(Imino)imidazolidine-Type Guanidines*

In 2000, Ishikawa's group prepared a variety of modified guanidines as potential chiral organobase catalysts based on 2-chloro-1,3-dimethylimidazolinium chloride (DMC) chemistry [1–3]. Among them, it was found that a 2-(imino)imidazolidine-type monocyclic guanidine incorporating a phenylalaninol moiety as an external hydrogen donor, 1,3-dimethyl-4,5-diphenyl-2-(1-hydroxymethyl-2-phenylethyl) iminoimidazolidine (Chiba-G, **1a**) (Fig. 1), was a respectable catalyst for the Michael addition of the benzophenone imine of glycinate **2a** with activated vinyl compounds **3** [4].

The Michael addition using methyl vinyl ketone (**3a**) as an acceptor in the presence of (+)-Chiba-G [(+)-**1a**] was performed in THF and afforded the (*R*)-adduct (*R*)-**4a** in high yield (90%) and with good enantioselectivity (96% *ee*). Satisfactory results (85% yield, 97% *ee*) were also achieved when using methyl acrylate (**3b**) as a Michael acceptor if the reaction was carried out without any solvent (Scheme 1).

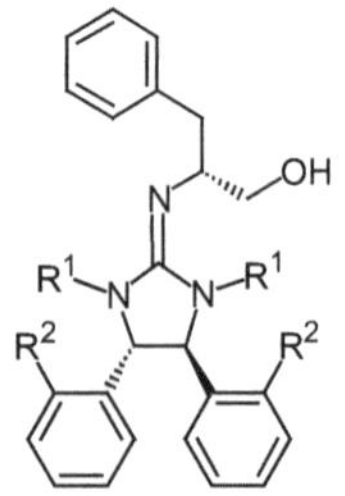


**1**

**a**: $R^1$ = Me, $R^2$ = H [(+)-Chiba-G]
**b**: $R^1$ = Bn, $R^2$ = H
**c**: $R^1$ = $R^2$ = Me

**Fig. 1** (+)-Chiba-G [(+)-**1a**] and related 2-(imino) imidazolidine-type guanidines (**1b** and **1c**)

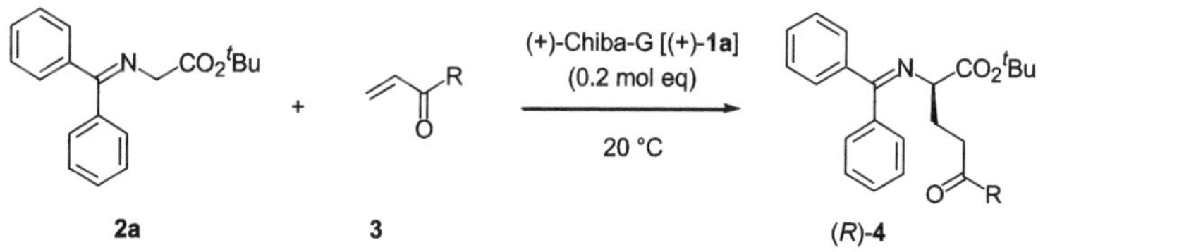


**Scheme 1** (+)-Chiba-G [(+)-**1**]-catalyzed Michael additions of the benzophenone-glycinate Schiff's base **2a** with activated olefins **3**

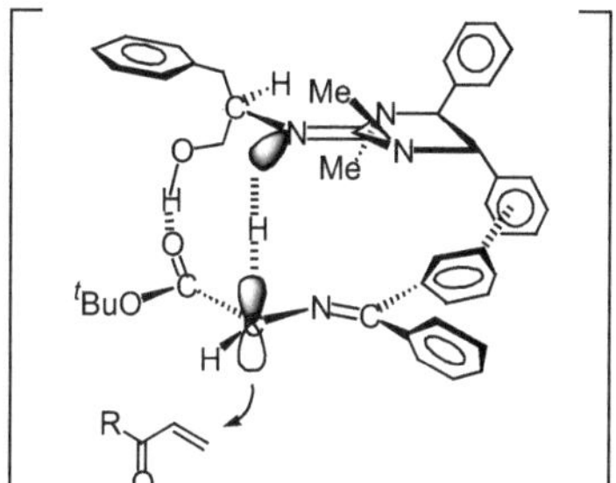


**Fig. 2** A plausible transition state **5** for (+)-Chiba-G [(+)-**1a**]-catalyzed Michael additions of the benzophenone-glycinate Schiff's base **2a** with activated olefins **3**

In order to explain the good asymmetric induction, a plausible transition state **5** was proposed, in which the benzophenone-glycinate Schiff's base **2a** is locked to Chiba-G (**1a**) on both ends of the molecule through a hydrogen bond to the ester carbonyl group and a CH-π interaction between the aromatic rings. This results in the exclusive deprotonation of the pro-*S* hydrogen in the substrate by the external guanidinyl nitrogen of the catalyst, followed by the addition of the resulting anion to the Michael acceptor from the opposite *re*face (Fig. 2) [4].

This Chiba-G (**1a**)-catalyzed Michael addition reaction was successfully applied to the enantioselective synthesis of the nicotine skeleton using pyridinyl vinyl ketone as a Michael acceptor. An excellent enantioselection of 94% *ee* was achieved [5].

Kobayashi's group applied Chiba-G (**1a**) to the Mannich reaction of the fluorenone-derived glycinate Schiff's base **2b** with *N*-Boc-protected aldimines **6** and observed the effective production of the adducts **7** with high *syn*-selectivities (up to >99/1) and enantioselectivities (up to 98% *ee*) (Scheme 2) [6]. The fluorenone imine moiety of the product was readily hydrolyzed under mildly acidic conditions. Thus, treatment of **7a** (R=Ph) with 1 N HCl/THF at 0°C for 1 h afforded α,β-diamino ester **8a** in 95% yield as a hydrochloride salt.

Structural modifications of Chiba-G (**1a**) indicated that not only the presence of the phenylalaninol moiety on the external guanidinyl nitrogen but also the relative configurations of the three chiral centers in Chiba-G (**1a**) were crucial for the

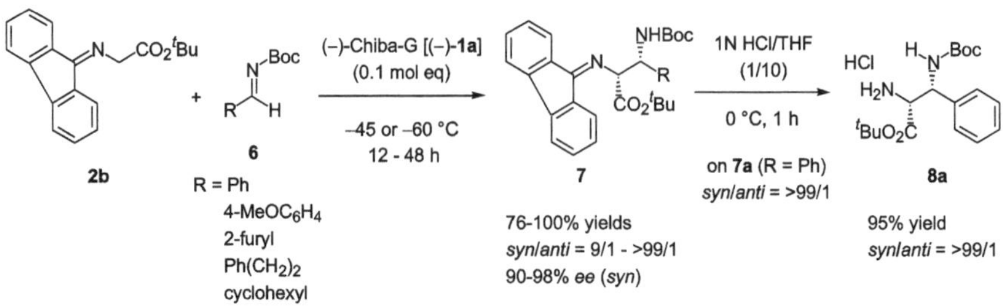


**Scheme 2** (−)-Chiba-G [(−)-**1a**]-catalyzed Mannich reaction of the fluorenone-derived glycinate Schiff's base **2b** with *N*-Boc aldimines **6** and the following acid hydrolysis of the adduct **7a**

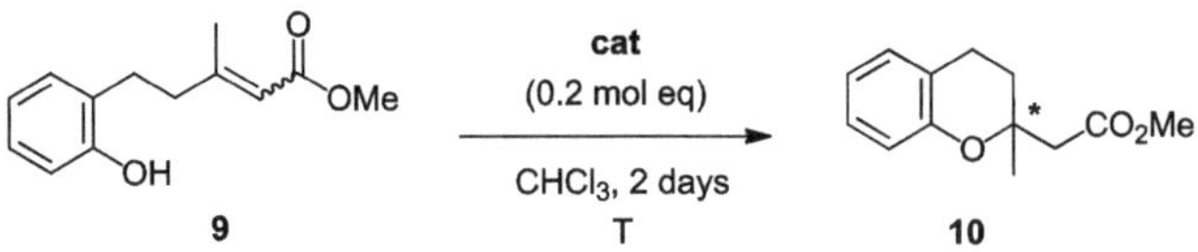


| Substrate | cat. | T | product (10) yield (%) | ee (%) | (config.) |
|---|---|---|---|---|---|
| (*E*)-**9** | (−)-**1a** | rt | 58 | 32 | (*S*) |
| (*Z*)-**9** | (−)-**1a** | rt | 75 | 71 | (*R*) |
| (*Z*)-**9** | (−)-**1c** | rt | 83 | 76 | (*R*) |
| (*Z*)-**9** | (−)-**1c** | 0 °C | 41 | 80 | (*R*) |

**Scheme 3** Enantioselective chromane cyclization by intramolecular oxo-Michael additions of phenol acrylates **9**

catalytic activity [4]. On the other hand, displacement of the *N*-methyl groups to more sterically demanding benzyl (Bn) groups (**1b**) or changing the phenyl to 2-methylphenyl pendants (**1c**) (Fig. 1) resulted in almost no improvement [7].

Chiba-G (**1a**) also worked as an effective catalyst in the Michael addition of 2-cyclopenten-1-one and dibenzyl malonate [8] and in the constructions of chromane skeletons **10** by the 6-*exo*-trig intramolecular oxa-Michael cyclization reaction of phenol acrylates **9** (Scheme 3) [9]. In this case, the (*E*/*Z*)-geometry of the acrylate unit in **9** was quite important for the asymmetric induction at the quaternary carbon center, and guanidine **1c**, carrying 2-methylphenyl pendants, was found to be the most effective catalyst. Thus, treatment of the (*Z*)-substrate (*Z*)-**9** with (−)-**1c** at room temperature (rt) afforded (*R*)-chromane (*R*)-**10** in 83% yield with 76% *ee*. Performing the reaction at 0°C caused to increment the *ee* to 80%, but the yield was lowered to 41%. The same 6-*exo*-trig intramolecular oxa-Michael cyclization strategy was applied to the enantioselective construction of the carbon framework in vitamin K [10].

Solid-state catalysts are considered as eco-friendly, sustainable synthetic tools because of their ease of handling, recyclability, and reduction of waste. Ishikawa's group examined the linking of Chiba-G (**1a**) to polymeric supports for the preparation of solid-state chiral guanidine catalysts (Fig. 3). Guanidine **11**, linked to

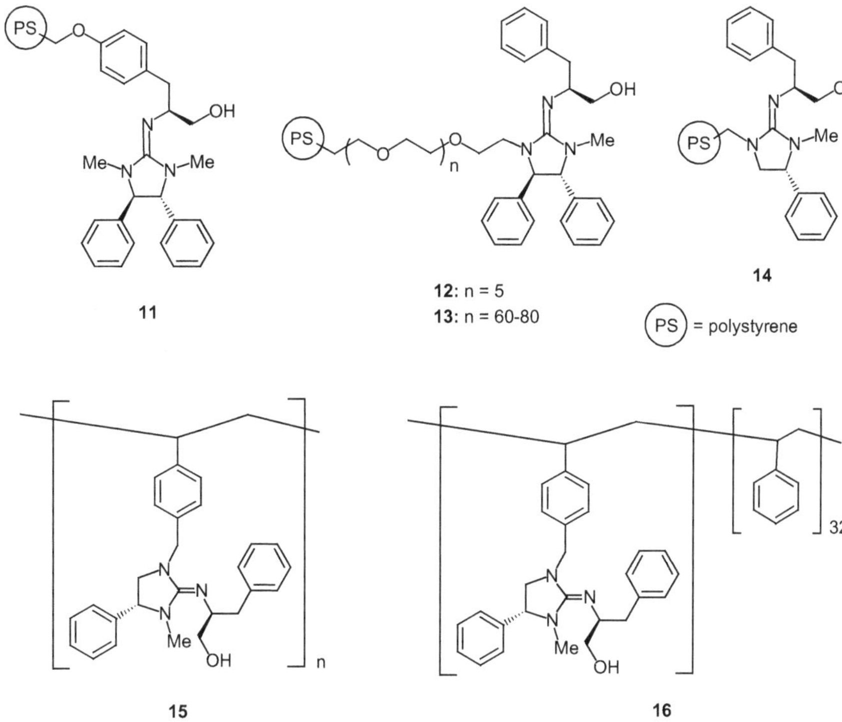


**Fig. 3** Solid-state 2-(imino)imidazolidine-type guanidine catalysts **11–16**

polystyrene through the phenyl ring of the benzyl group on the external guanidinyl nitrogen substituent, was identified as a suitable replacement for Chiba-G (**1a**) in the Michael addition reaction of the benzophenone-glycinate Schiff's base **2a** with methyl vinyl ketone (**3a**) [11]. On the other hand, guanidine catalysts **12** and **13**, in which an internal nitrogen atom was bound to polystyrene, showed a significant decrease in reactivity. The same tendency was observed in trials using the related solid-state guanidine catalysts **14–16**. Only moderate selectivities were observed when the polymeric guanidines **15** and **16** were used [12].

## *2.2 $C_2$-Symmetric Guanidine Hydroxides and Their Analogs*

In 1999, Murphy's group prepared $C_2$-symmetric [4.4.0]bicyclic guanidines (**17**) carrying a hydroxymethyl-substituted *spiro*-heteroketal functionality derived from (*S*)-malic acid (Fig. 4a). The guanidine hydroxide **17a** was screened as a free base catalyst in the Henry reaction of nitromethane and isovaleraldehyde [13] and as a phase-transfer catalyst in the benzylation of the benzophenone-glycinate Schiff's

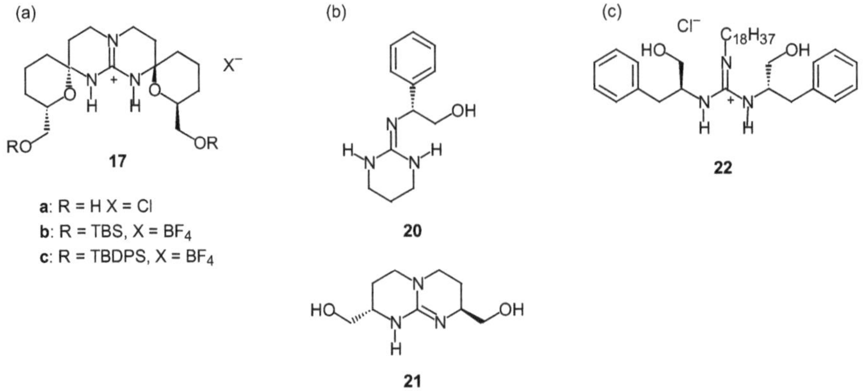


**Fig. 4** $C_2$-Symmetric guanidine hydroxide catalysts and their analogs. (**a**) Murphy's guanidines. (**b**) Taylor's guanidines. (**c**) Nagasawa's guanidine

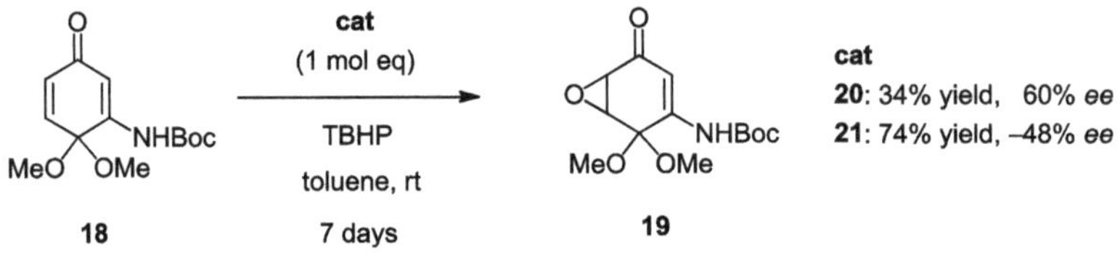


**Scheme 4** Nucleophilic epoxidations of amidequinone methide **18** catalyzed with cyclic guanidine hydroxides **20** and **21**

base **2a** [14]. However, only low asymmetric inductions were observed. In the latter reaction, the (*R*)-adduct was afforded in 15% yield with 21% *ee*, whereas the silyl-protected derivatives led to better results [TBS derivative **17b**: 70% yield with 65% *ee*; TBDPS derivative **17c**: 80% yield with 74% *ee*] under the same conditions. The low conversion and *ee* achieved with the guanidine hydroxide catalyst **17a** might be caused by its poor solubility in the organic phase of the reaction.

In 2003, Taylor's group prepared enantiopure guanidines with hydroxyl functions (Fig. 4b) and evaluated them in the nucleophilic epoxidation of amide quinone methide **18** using *t*-butylhydroperoxide (TBHP) [15]. When the monocyclic guanidine **20** was used as a catalyst, the (+)-epoxide (+)-**19** was obtained in 34% yield and 60% *ee* together with 41% of recovered starting material **18**. On the other hand, the use of $C_2$-symmetric [4.4.0]bicyclic guanidine **21** provided the (−)-epoxide (−)-**19** in 74% yield and 48% *ee* (Scheme 4). No enantioselections were observed in the reactions catalyzed with the corresponding hydroxyl-protected analogs.

The monocyclic guanidine **20** was also found to mediate the epoxidation of *trans*-chalcone (71% yield, 15% *ee*) and 2-methylnaphthoquinone (68% yield, 31% *ee*) albeit with only moderate enantioselection [16].

In 2008, Nagasawa's group designed a $C_2$-symmetric acyclic guanidine **22** incorporating two phenylalaninol moieties (Fig. 4c) [17]. After optimization of the reaction conditions, application as a phase-transfer catalyst for the nucleophilic epoxidation of a series of chalcones led to acceptable conversions (56–99%) and enantioselectivities in the range of 60–73% *ee*. The hydroxyl groups were crucial for asymmetric induction, as their protection with methyl groups resulted in the loss of enantioselectivity.

## 2.3 [4.3.0]Bicyclic Guanidine Hydroxides Carrying a Diarylprolinol-Like Structure

In 2010, Sugimura's group introduced the [4.3.0]bicyclic guanidine hydroxide **23** carrying a partial diarylprolinol-like structure as an effective organocatalyst for the stereoselective construction of the chiral quaternary carbon atom of 5*H*-oxazol-4-ones **24** by asymmetric aldol reactions with aldehydes **25** [18] or Michael additions to alkynyl carbonyl compounds **27** (Scheme 5) [19]. The aldol reactions showed a broad tolerance of functionalities. Various combinations of 5*H*-oxazolones **24** and

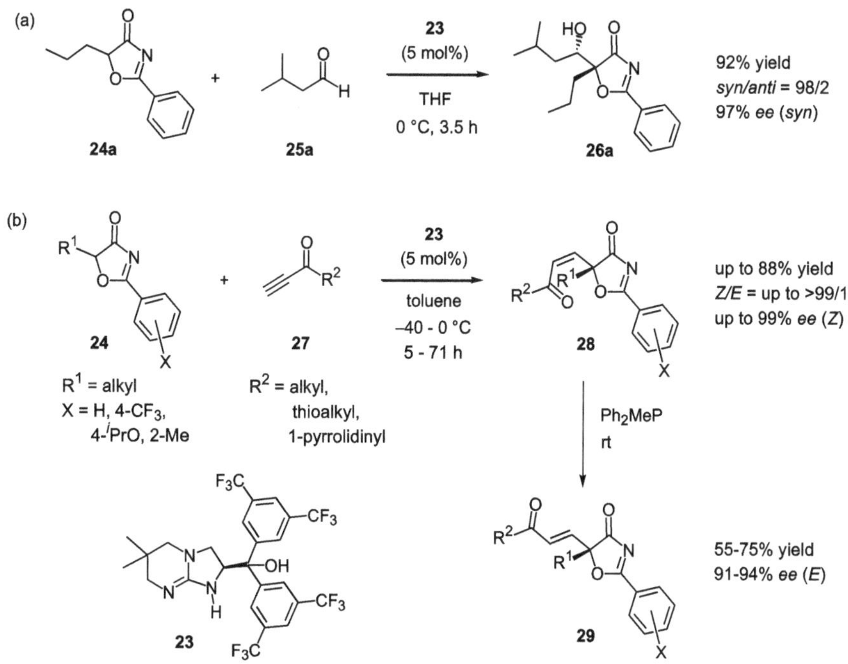


**Scheme 5** [4.3.0]Bicyclic guanidine hydroxide-catalyzed asymmetric reactions of 5*H*-oxazol-4-ones **24**: (**a**) aldol reactions with aldehydes **25** and (**b**) Michael additions with alkynyl carbonyl compounds **27**

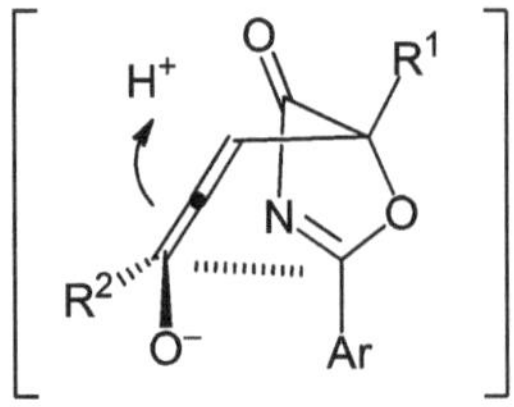
30

**Fig. 5** Supposed transition state **30** for the Michael additions of 5*H*-oxazol-4-ones **24** and alkynyl carbonyl compounds **27**

aldehydes **25** afforded the products **26** in moderate to good yields (43–92%), with moderate to excellent *syn*/*anti* selectivities (67/33 to >98/2) and with high enantioselection of the major *syn*-product (92–97% *ee*) [18]. A typical reaction of oxazolone **24a** and isovaleraldehyde (**25a**) is shown in Scheme 5a, in which the adduct **26a** was obtained in 92% yield, with 98/2 *syn*/*anti* selectivity, and with 97% *ee* for the major *syn*-isomer.

In the Michael addition reactions, high enantiomeric and diastereomeric controls leading to highly (*Z*)-selective 1,4-addition reactions (up to 99% *ee* and *Z*/*E* = >99/1) were also achieved (Scheme 5b) [19]. The (*Z*)-selectivity was explained as shown in Fig. 5: One face of the enolate anion in the intermediate **30** is shielded by the 5*H*-oxazolone ring, and therefore, protonation occurred from the other face giving the thermodynamically unstable (*Z*)-isomer **28**.

Isomerization of the (*Z*)-product **28** to the (*E*)-isomer **29** was achieved by aid of diphenylmethylphosphine ($Ph_2MeP$) without loss of optical purity [20]. Thus, reactions of oxazol-4-ones **24** and alkynones **27** in toluene in the presence of catalyst **23** at 0°C followed by treatment of the reaction mixture with $Ph_2MeP$ at room temperature afforded the adducts **29** with (*E*)-configuration in 55–75% yield and with 91–94% *ee*.

## 2.4 Tartrate-Derived Guanidines

Tartrate-derived guanidines **31** were screened as organocatalysts for the Michael additions of 3-substituted oxindoles **32** and nitroolefins **33** by Wang's group [21]. 3,5-di(*tert*-Butyl)aniline-derived guanidine **31a** afforded the 3,3-disubstituted oxindole **34a** in 72% yield with good diastereoselectivity (95/5) and 77% *ee* for the major isomer when the oxindole **32a** was treated with nitrostyrene (**33a**) in toluene at 0°C for 16 h (Scheme 6). On the other hand, only a modest diastereoisomeric ratio (66/34) was observed by the application of the aminoalcohol-derived guanidine **31b** under the same conditions. It is interesting to note, however, that a reversal in the sense of enantioselectivity for the major diastereoisomer occurred.

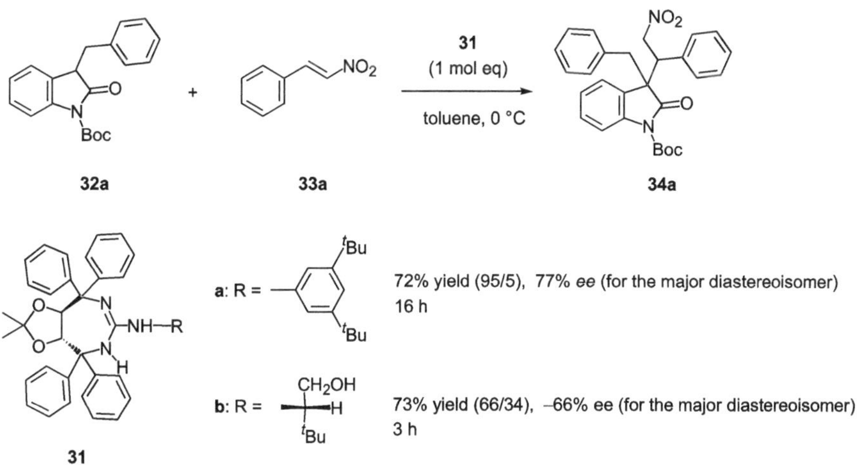


**Scheme 6** Representative example for the Michael addition of oxindoles **32** and nitroolefins **33** catalyzed with tartrate-derived guanidines **31**

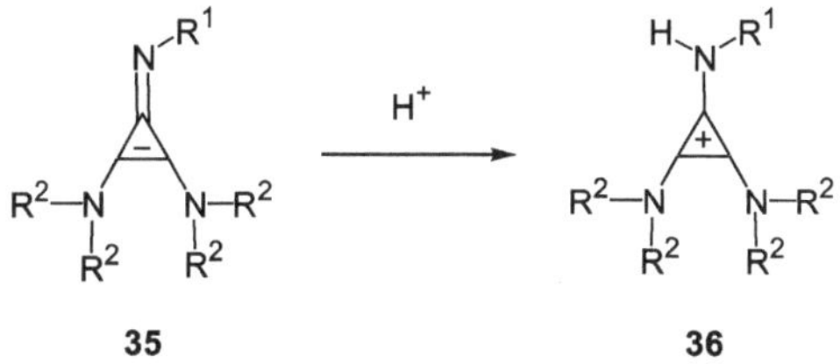


**Fig. 6** Cyclopropenimines **35** as strong Brønsted bases

## 2.5 *Cyclopropenimine-Type Quasi-Guanidine Bases*

The 2π-electron cyclopropenium cation provides substantial aromatic resonance stabilization to the conjugated acids of cyclopropenimines. In comparison with the analogous guanidines, this additional stabilization renders 2,3-bis(dialkylamino) cyclopropenimines **35** highly basic. The conjugated acids **36** are stabilized by three nitrogen lone pairs plus the additional aromatic cyclopropenium cation (Fig. 6). Lambert's group measured the acidity ($pK_{BH+} = 26.9$) of the conjugate acid of cyclopropenimine **35** ($R^1 = {}^t$Bu, $R^2 = {}^i$Pr) in acetonitrile to be comparable with superbases such as the bicyclic guanidine 1,5,7-triazabicyclo [4.4.0]dec-5-ene (TBD) ($pK_{BH+} = 26.03$) and the phosphazene base $P_1$-$^t$Bu ($pK_{BH+} = 26.98$) [22].

The Michael addition of the benzophenone-glycinate Schiff's base **2a** with methyl acrylate (**3b**) was investigated using cyclopropenimine catalysts **37** as a forum for comparison to Chiba-G (**1a**). The presence of a hydroxyl group was found to be crucial for both reactivity and enantioselectivity (Scheme 7). Sterically demanding dialkylamino substituents at the 2- and 3-positions were also found to

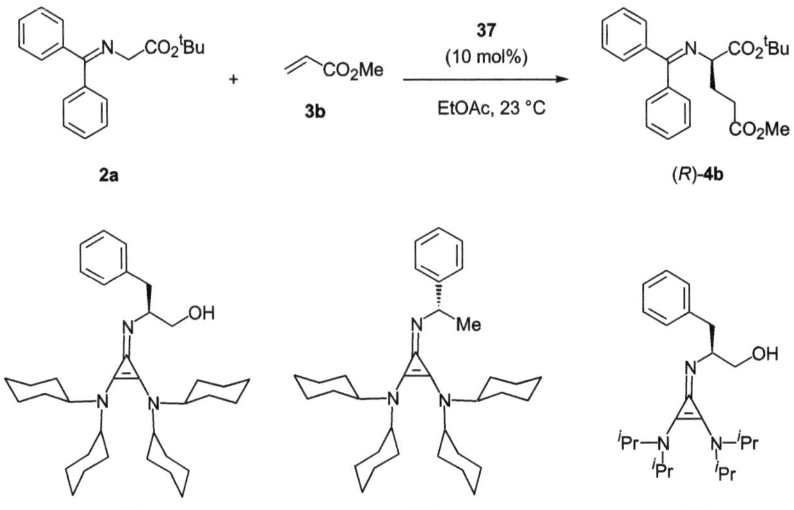


**Scheme 7** Cyclopropenimine **37**-catalyzed Michael additions of the benzophenone-glycinate Schiff's base **2a** with methyl acrylate (**3b**)

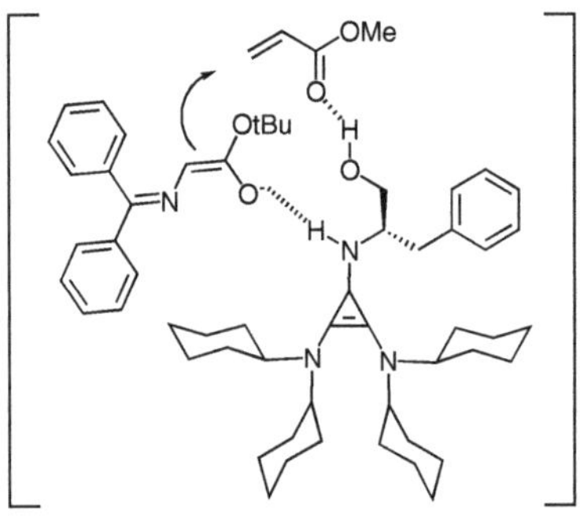


**Fig. 7** Stereochemical rationale and proposed transition state **38** for the cyclopropenimine **37a**-catalyzed enantioselective conjugate addition

be important for optimal performance. Thus, the dicyclohexylamine-derived catalyst **37a** incorporating the same phenylalaninol moiety as in Chiba-G (**1a**) was the most promising catalyst. The Michael addition method was applicable to a preparative scale, and 25 g of the product **4b** (97% yield with 99% *ee*) was obtained within 8 h using 2.5 mol% of catalyst.

A tentative mechanistic and stereochemical rationale for this transformation was proposed in a transition state **38** based on the X-ray structure of the protonated catalyst (Fig. 7).

# 3 Guanidine–Amide Catalysts

In 2009, Feng's group designed bifunctional guanidines featuring chiral, rigid, and cyclic α-amino amide backbones derived from L-proline, L-pipecolic acid, or L-ramipril acid, for the asymmetric Michael addition of β-keto esters **40** to nitroolefins **33**, exploiting a dual mode of activation in one molecule of catalyst. The pipecolic acid-derived catalyst **39a** demonstrated high stereoselectivities [up to >99/1 *syn*/*anti* selectivity and 97% *ee* (*syn*)] and yields up to 99% for a wide range of substrates [23]. The reaction of cyclopentanone-2-carboxylate **40a** and 4-bromo-nitrostyrene (**33b**) is shown in Scheme 8 as a representative example. Adduct **41a** was obtained in 99% yield and with high stereoselectivities [*syn*/*anti* = 99/1 and 95% *ee* (*syn*)].

The comparative experiment employing the *N*-methylated catalyst derivative **39b** as well as X-ray diffraction analysis of the catalyst revealed that both the guanidine group and the N–H proton of the amide are important for successful dual activation. A plausible transition state **42** is shown in Fig. 8, in which both substrates are simultaneously activated by hydrogen bonding.

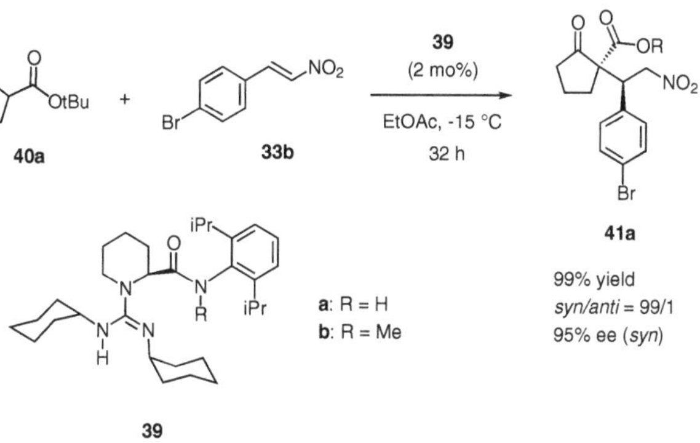


**Scheme 8** Representative example for the Michael addition of β-keto esters **40** and nitroolefins **33** catalyzed by guanidine–amide **39a**

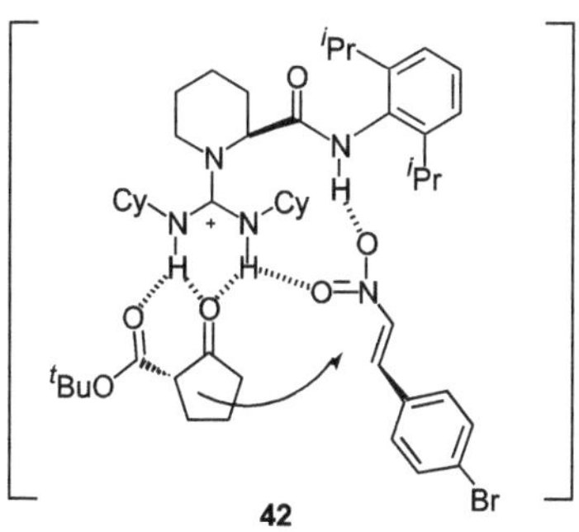


**Fig. 8** Dual activation in the transition state **42** for the guanidine–amide **39a**-catalyzed Michael addition of cyclopentanone-2-carboxylate **40a** and 4-bromo-nitrostyrene (**33b**)

Feng's group further designed a 1,2-diphenylethylene-1,2-diamine-linked $C_2$-symmetric chiral bisguanidine **43** containing two pipecolic acid moieties as a highly efficient catalyst in the inverse electron demand hetero-Diels–Alder (IEDHDA) reaction of chalcones **44** with azlactones **45** [24]. A wide variety of γ,δ-unsaturated δ-lactone derivatives **46** with α-quaternary-β-tertiary stereocenters was obtained as single diastereoisomers in high yields (up to 88%) and with excellent enantioselectivities (up to 99% *ee*), along with small amounts of Michael addition products. The reaction of chalcone (**44a**) and azlactone **45a** is shown in Scheme 9 as a representative example, in which the adduct **46a** was obtained in 73% yield and 96% *ee*.

The chiral guanidine catalyst was recovered from the reaction mixture and reused without any loss of catalytic activity. Hydrogen bonds were considered to be crucial for the activation and stereoselection in this reaction. In the proposed bifunctionally activated transition state **47**, the N–H moiety of the amide could act as a Brønsted acid to activate the chalcone by lowering its LUMO energy through a hydrogen bond. The enolized azlactone was recognized by the guanidine moiety, associating with the N–H proton of the amide on the other side via hydrogen bonds, leading to (3*S*,4*R*)-products by attack from the *re*face of the chalcone (Scheme 9).

The related chiral mono-guanidine **48** catalyzed the domino reactions of *o*-hydroxy aromatic aldimines **49** and azlactones **45** to afford *cis*-3,4-diaminochroman-2-ones *cis*-**50** in high yields (up to 99%) with excellent stereoselectivities [up to *cis*/*trans* = >99/1 and 96% *ee* (*cis*)] [25]. Interestingly, the

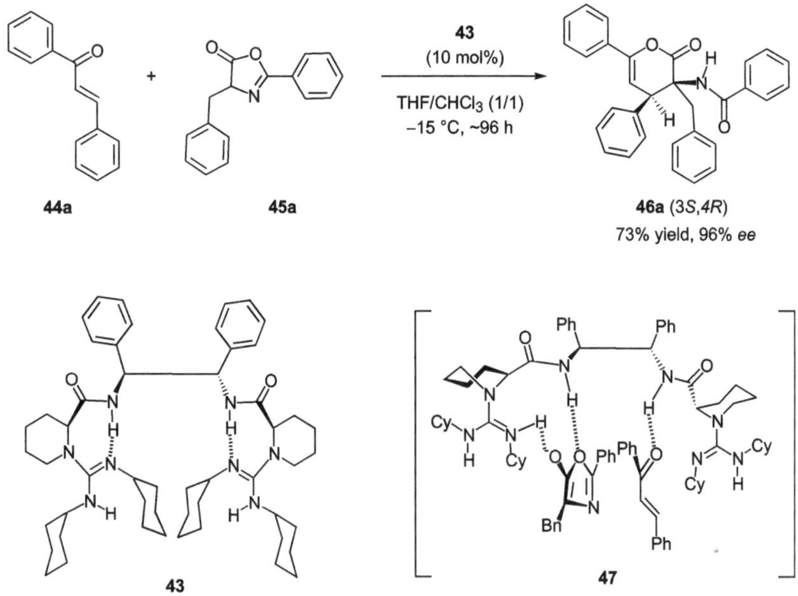


**Scheme 9** The guanidine–amide **43**-catalyzed IEDHDA reaction of chalcone (**44a**) with azlactone **45a** and its proposed transition state **47**

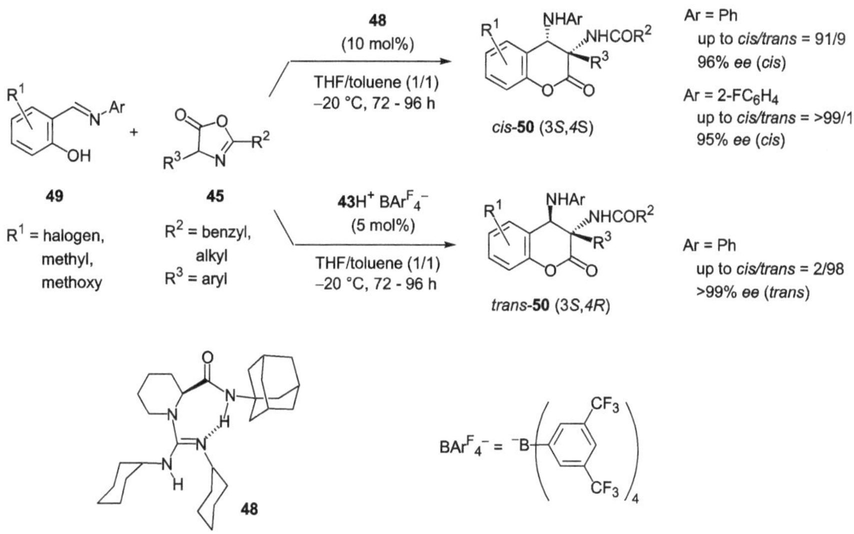


**Scheme 10** Guanidine–amide-catalyzed domino reactions of *o*-hydroxy aromatic aldimines **49** and azlactones **45**

*trans*-isomers *trans*-**50** were obtained as the major products (d.r. up to 98/2) when the diphenylethylenediamine-bridged $C_2$-symmetric bisguanidine **43** was employed as a catalyst in its salt form (**43**$H^+$ $BAr^F{}_4{}^-$) (Scheme 10).

The latter bis-guanidinium salt catalyst **43**$H^+$ $BAr^F{}_4{}^-$ was also found to be effective in the enantioselective intramolecular aza-Michael addition as well as one-pot bromination reaction of 2-(*N*-tosylanthranyl)acrylates **51**, providing 2-substituted dihydroisoquinolines **52** (up to 99% yield and 99% *ee*) and brominated dihydroisoquinolines *trans*-**53** [up to 95% yield, *trans*/*cis* = 96/4, and 95% *ee* (*trans*)], respectively (Scheme 11) [26]. The major *trans*-product *trans*-**53a** (R=Ph, *trans*/*cis* = 91/9, 93% *ee*) was smoothly isomerized into the *cis*-isomer (*cis*-**53a**) (*cis*/*trans* = 95/5, 93% *ee*) without any loss of enantiopurity by treatment with potassium hydroxide in toluene for 1 h. The inversion of configuration proceeds through an $S_E$ 1 mechanism with the *cis*-isomers *cis*-**53** being the thermodynamically more stable products.

An alternative phenylene-linked $C_2$-symmetric guanidine–amide catalyst **54** was developed for Mannich-type reactions of α-isothiocyanato imide **55** with *N*-tosyl-protected aldimines **56** [27]. The reactions were performed in the presence of 4-cyanobenzoic acid as an additive and were suitable for a broad substrate scope. Optically active α,β-diamino acid derivatives **57** were obtained in high yields (up to 99%), with good diastereoselectivities (up to *trans*/*cis* = >95/5) and excellent enantioselectivities [up to >99% *ee* (*trans*)] (Scheme 12). A possible transition state was proposed, in which both substrates were effectively activated by the bisguanidine catalyst through hydrogen bonds.

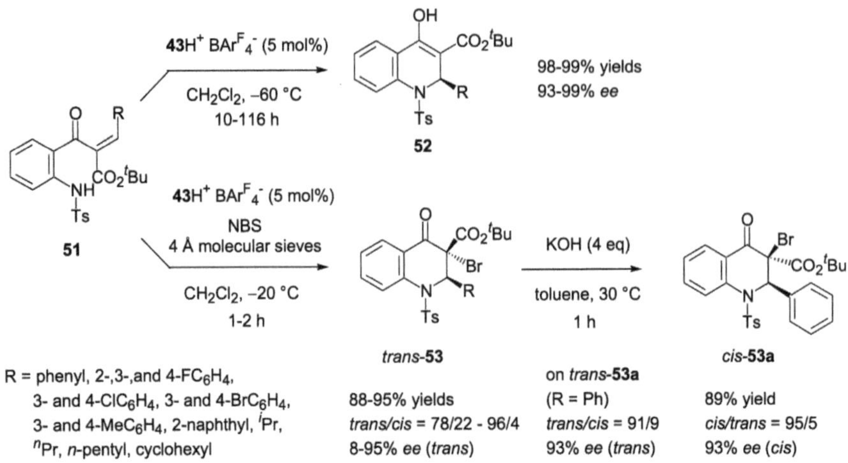


**Scheme 11** Guanidine–amide-catalyzed intramolecular aza-Michael additions and one-pot bromination reactions of 2-(*N*-tosylanthranyl)acrylates **51**

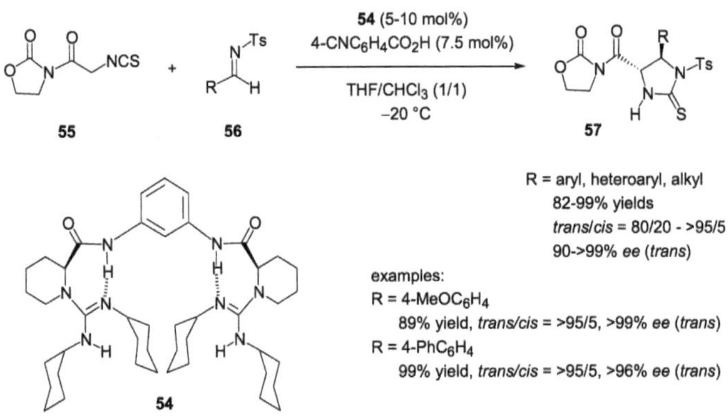


**Scheme 12** Guanidine–amide **54**-catalyzed Mannich-type reactions of α-isothiocyanato imide **55** with *N*-tosyl-protected aldimines **56**

Catalyst **54** was further elaborated into the sophisticated structure **58a**, in which the 1,2-diphenylethylene-1,2-diamine linker is connected to ramipril acid ([3.3.0] azabicyclooctane framework) instead of pipecolic acid (piperidine framework) derivates. This catalyst was used for the direct vinylogous Michael addition of α,β-unsaturated γ-butyrolactam **59** with alkylidene malonates **60** (Scheme 13) [28]. The reaction is suitable for a wide range of substrates, and the adducts **61** were generally obtained in reasonable yields (up to 93%) and high selectivities [up to 95/5 diastereoselectivity and 94% *ee* (for the major isomer)]. Comparative experiments with the corresponding *N*-Me-protected guanidine derivative **58b** showed both decreased reactivity (15% yield) and reduced stereoselectivity for

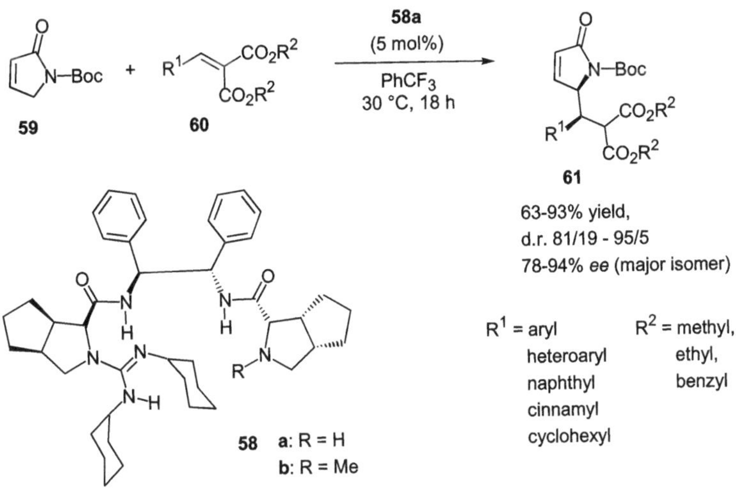


**Scheme 13** The guanidine–amide (**58a**)-catalyzed vinylogous Michael additions of α,-β-unsaturated γ-butyrolactam (**59**) and alkylidene malonates (**60**)

the major isomer (71% *ee*). A possible activation model for the reaction was also proposed.

A similar guanidine-amide catalyst containing an *N*-(diphenylmethyl) prolinamide skeleton was introduced as an organocatalyst for the Michael additions of arylacetylsilanes with 1-aryl-(or heteroaryl)-2-nitroethenes by Tang's group [29]. Reactions were carried out with 0.1 mmol acylsilane, 15 mol% catalyst, and 60 mg 4 Å molecular sieves in toluene (1.5 mL) at 0°C for 12 h, with 0.25 mmol nitroolefin dissolved in toluene added to the mixture in 4 portions over 8 h. The α,β-disubstituted γ-nitrobutyrylsilane adducts were afforded in 61–87% yield with *syn*/*anti* = 12/1 to >99/1 and 89–97% *ee* for the major *syn*-diastereomers. Thus, {[β-(4-bromophenyl)-γ-nitro-α-phenyl]butyryl}dimethyl-phenylsilane was isolated as an adduct in 77% yield, *syn*/*anti* = >99/1, and 96% *ee* (*syn*).

## 4 Guanidine–Amine Catalysts

The guanidine–amine **62** was prepared by mono-guanidinylation of (1*S*,2*S*)-cyclohexane-1,2-diamine and used as a chiral organocatalyst for the enantioselective conjugated additions of α,α-disubstituted aldehydes **63** to maleimides **64** in aqueous DMF at 25°C (Scheme 14) [30]. The (*R*)-adducts (*R*)-**65** were generally obtained in high to quantitative yields (84–99%) and with good enantioselections (up to 93% *ee*). A different transition state as compared to the reactions promoted with amine–thiourea catalysts was proposed because opposite asymmetric inductions were observed in the Michael adducts.

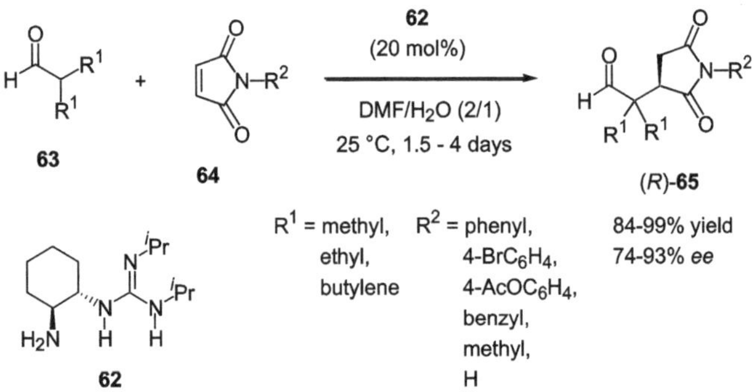


**Scheme 14** Guanidine–amine **62**-catalyzed Michael additions of α,α-disubstituted aldehydes **63** to maleimides **64**

The guanidine–amine catalyst **62** was also applicable to conjugate additions using arylated and heteroarylated nitroalkenes as Michael acceptors in the presence of imidazole as an additive, giving adducts in high yields (70–95%) and enantioselectivities (65–80% *ee*) [31]. The observed sense of stereoinduction was approached using theoretical calculations.

# 5 Conclusion and Outlook

Bifunctional guanidines carrying active hydrogens such as hydroxyl, amide, and amine groups at appropriate positions satisfactorily acted as efficient organocatalysts in enantioselective syntheses. Hydrogen bondings between substrates and the guanidine catalyst should be responsible for the stereochemical constraints in the transition states leading to high asymmetric inductions. Solid-state catalysts are considered as eco-friendly, sustainable synthetic tools because of their ease of handling, recyclability, and reduction of waste. However, only a single polymer-bound imidazolidine-type guanidine hydroxide based on Chiba-G was successfully introduced as a solid-state catalyst to date. Therefore, it should be emphasized that other types of solid-state bifunctional guanidine catalysts applicable to reactions under solvent-free conditions should be explored for the extensive development of sustainable organic chemistry in the future.

# References

1. Isobe T, Fukuda K, Tokunaga T, Ishikawa T (2000) J Org Chem 65:7770–7773
2. Isobe T, Fukuda K, Tokunaga T, Seki H, Yamaguchi K, Ishikawa T (2000) J Org Chem 65:7774–7778

3. Isobe T, Fukuda K, Yamaguchi K, Seki H, Tokunaga T, Ishikawa T (2000) J Org Chem 65:7779–7785
4. Ishikawa T, Araki Y, Kumamoto T, Seki H, Fukuda K, Isobe T (2001) Chem Commun 245–246
5. Zhang G, Kumamoto T, Heima T, Ishikawa T (2010) Tetrahedron Lett 51:3927–3930
6. Kobayashi S, Yazaki R, Seki K, Yamashita Y (2008) Angew Chem Int Ed 47:5613–5615
7. Ryoda A, Yajima N, Haga T, Kumamoto T, Nakanishi W, Kwahata M, Yamaguchi K, Ishikawa T (2008) J Org Chem 73:133–141
8. Kumamoto T, Ebine K, Endo M, Araki Y, Fushimi Y, Miyamoto I, Ishikawa T, Isobe T, Fukuda K (2005) Heterocycles 66:347–359
9. Saito N, Ryoda A, Nakanishi W, Kumamoto T, Ishikawa T (2008) Eur J Org Chem 2759–2766
10. Tokunou S, Nakanishi W, Kagawa N, Kumamoto T, Ishikawa T (2012) Heterocycles 84:1045–1056
11. Ishikawa T, Heima T, Yoshida M, Kumamoto T (2014) Helv Chim Acta 97:307–314
12. Disadee W, Ishikawa T (2005) Mol Divers 1–11
13. Howard-Jones A, Murohy PJ, Thomas DA (1999) J Org Chem 64:1039–1041
14. Allingham MT, Howard-Jones A, Murphy PJ, Thomas DA, Caulkett PWR (2003) Tetrahedron Lett 44:8677–8680
15. McManus JC, Carey JS, Taylor RJK (2003) Synlett 365–368
16. McManus JC, Genski T, Carey JS, Taylor RJK (2003) Synlett 369–371
17. Shin B, Tanaka S, Kita T, Hashimoto Y, Nagasawa K (2008) Heterocyles 76:801–810
18. Misaki T, Takimoto G, Sugimura T (2010) J Am Chem Soc 132:6286–6287
19. Misaki T, Kawano K, Sugimura T (2011) J Am Chem Soc 133:5695–5697
20. Misaki T, Jin N, Kawano K, Sugimura T (2012) Chem Lett 41:1675–1677
21. Zou L, Bao X, Ma Y, Song Y, Qu J, Wang B (2014) Chem Commun 50:5760–5762
22. Bandar JS, Lambert TH (2012) J Am Chem Soc 134:5552–5555
23. Yu Z, Liu X, Zhou L, Lin L, Feng X (2009) Angew Chem Int Ed 48:5195–5198
24. Dong S, Liu X, Chen X, Mei F, Zhang Y, Gao B, Lin L, Feng X (2010) J Am Chem Soc 132:10650–10651
25. Dong S, Liu X, Zhang Y, Lin L, Feng X (2011) Org Lett 13:5060–5063
26. Xiao X, Liu X, Dong S, Cai Y, Lin L, Feng X (2012) Chem Eur J 18:15922–15926
27. Chen X, Dong S, Qiao Z, Zhu Y, Xie M, Lin L, Liu X, Feng X (2011) Chem Eur J 17:2583–2586
28. Yang Y, Dong S, Liu X, Lin L, Feng X (2012) Chem Commun 48:5040–5042
29. Wu L, Li G, Fu Q, Yu L, Tang Z (2013) Org Biomol Chem 11:443–447
30. Avila A, Chinchilla R, Nájera C (2012) Tetrahedron Asymmetry 23:1625–1627
31. Avila A, Chinchilla R, Fiser B, Gómez-Bengoa E, Nájera C (2014) Tetrahedron Asymmetry 25:462–467

Top Heterocycl Chem (2017) 50: 71–94
DOI: 10.1007/7081_2015_155

Published online: 27 June 2015

# Bifunctional Guanidines as Hydrogen-Bond-Donating Catalysts

**Yusuke Kobayashi and Yoshiji Takemoto**

**Abstract** Bicyclic guanidines, bearing 2-aminoquinazolin-4-one and 3-aminobenzothiadiazine-1,1-dioxide core structures, work as bifunctional amine catalysts, due to a double hydrogen-bond-donating ability of two guanidine N-H protons, which are capable of activating both nucleophile and electrophile simultaneously. In fact, these guanidine catalysts revealed to promote several asymmetric reactions such as hydrazination of active methylene compounds with azodicarboxylate and 1,3-proton migration of alkynoates to allenoates as well as the oxa-Michael addition and epoxidation of α,β-unsaturated amides and esters with better chemical yields and higher enantioselectivities than the corresponding thioureas. The catalytic mechanisms of these asymmetric reactions and their synthetic applications for biologically active molecules are also discussed in this chapter.

**Keywords** Allene • Asymmetric reaction • Atorvastatin • Benzothiadiazine • Chromane • Dihydrobenzofuran • Epoxidation • Hydrazination • Hydrogen-bond donor • Isomerization • Organocatalysis • Oxa-Michael addition • Quinazoline • Thiourea • Total synthesis

## Contents



Y. Kobayashi and Y. Takemoto (✉)
Graduate School of Pharmaceutical Sciences, Kyoto University, Sakyo-ku, Kyoto 606-8501, Japan
e-mail: takemoto@pharm.kyoto-u.ac.jp

## Abbreviations

| | |
|---|---|
| DFT | Density functional theory |
| DIBAL | Diisobutylaluminum hydride |
| DMSO | Dimethylsulfoxide |
| ee | Enantiomeric excess |
| HB | Hydrogen bond |
| HPLC | High performance liquid chromatography |
| HWE | Horner-Wadsworth-Emmons |
| NMR | Nuclear magnetic rcsonance |
| PCM | Polarizable continuum model |
| TBHP | *tert*-Butylhydroperoxide |

# 1 Introduction

Guanidines are widely used as versatile organic bases in base-catalyzed reactions, because their $pK_b$ values can be tuned at will by changing the steric and electronic effects of the substituents on each three guanidine nitrogens, RN=C(NR′R″) NHR‴. On the other hand, the conjugate acids of guanidines, which are generated by the reaction with appropriate acids, enable to work as hydrogen-bond donor catalysts. Therefore, it would be reasonable to consider that the binary complexes or ion pairs consisting of guanidine derivatives and nucleophiles play a crucial role in a rate-determining or stereo-determining step. Consequently, guanidine catalysts have been extensively applied to a wide range of both acid- and base-catalyzed reactions including asymmetric reactions. Although, so far, a variety of chiral guanidines have been developed as Brønsted bases for a broad array of catalytic asymmetric reactions, there are only a few reports that guanidine derivatives, but not guanidinium salts, are directly employed as the hydrogen-bond-donating catalysts [1–7].

We recently discovered that bicyclic heterocycles such as 2-amino-4-oxoquinazoline and 3-aminobenzothiadiazine-1,1-dioxide display strong hydrogen-bond-donating ability, in some cases, more than a thiourea motif. In this chapter, the asymmetric reactions using bifunctional guanidine catalysts as well as the chemical structure and design concept of these catalysts are described [8–12].

# 2 Design of Bifunctional Guanidines

Hydrogen-bond (HB) donors are recognized as important and effective catalysts in organocatalysis. Thiourea is a representative motif, which enables to form a double hydrogen bonding interaction with an appropriate Lewis base such as chloride, cyanide, imine, ketone, and nitro compounds, due to the suitable positioning of two N-H protons [13–15]. We have already developed bifunctional aminothiourea/urea **1a/1b** and demonstrated that a wide range of asymmetric reactions including the Michael addition, the Mannich reaction, the aza-Henry reaction, and the Neber reaction were promoted by this catalyst to give the corresponding products with high enantioselectivity [16]. In these reactions, both the thiourea moiety and the tertiary amine of the catalyst cooperatively activate a nucleophile and an electrophile. However, the catalyst has the serious disadvantage to suffer from decomposition of the catalyst by the subjection to some kinds of oxidants and electrophiles due to the high reactivity of the sulfur atom [17]. To avoid this problem, we designed new hydrogen-bond donors bearing cyclic guanidine motifs, in which the thiourea moiety and the aryl group of the catalyst **1a** are linked as shown in Fig. 1 [8]. Furthermore, the acidities of two N-H protons of the guanidines can be enhanced by choosing a suitable tether (C=O or $SO_2$) as well as a substituent (R = F, $CF_3$, etc.) on the benzene ring.

From the synthetic point of view, we selected quinazoline- and benzothiadiazine-type catalysts **2** and **3**. Since guanidine catalysts possessing strong electron-withdrawing groups such as C=O and $SO_2$ are no longer expected to function as Brønsted bases, the tertiary amino group is essential to catalyze the reactions in the same manner as the bifunctional thiourea. These catalysts are readily synthesized in four steps from the corresponding anilines and anthranilic acids as shown in Scheme 1.

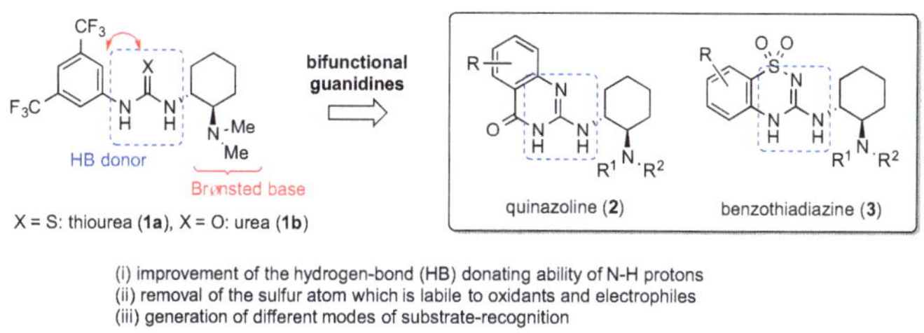


**Fig. 1** Design of bifunctional guanidine catalysts

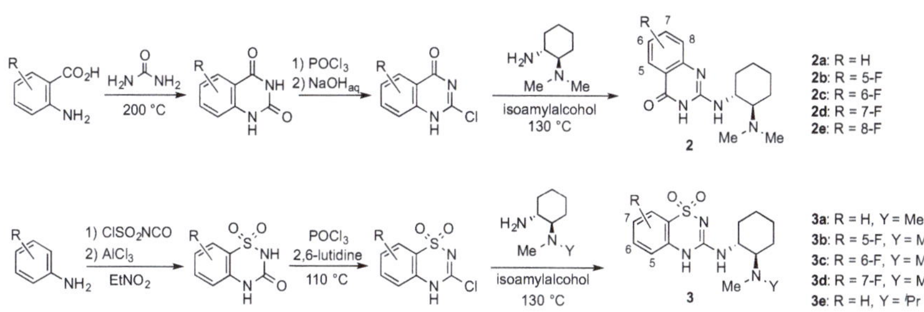


**Scheme 1** Preparation of bifunctional quinazolines **2** and benzothiadiazines **3**

To evaluate the HB-donating abilities of three bifunctional HB donors **1–3**, their association constants ($K_1$) were determined by the titration of an appropriate Lewis base to a solution of each HB donor (Scheme 2). After many trials, a chloride anion in acetonitrile revealed to be a suitable combination for the titration by means of UV spectroscopy rather than the more basic acetate anion in DMSO. In the latter case, an alternative acetate anion further interacted with the binary complex **A** (Cat·LB) to form the ternary complex **B** [Cat·(LB)$_2$], which causes a deviation from the equation of a one-step binding model. Under these conditions, the association constants of thiourea **1a**, quinazoline **2a**, and benzothiadiazine **3a** were estimated to be $1.2 \times 10^3$ ($\pm$37), $4.9 \times 10^2$ ($\pm$18), and $1.9 \times 10^3$ ($\pm$44), respectively. Therefore, the relative abilities of these HB donors to associate with Lewis bases were remarkably different and followed the order **3a** > **1a** > **2a** [8].

Since the two guanidine-type catalysts **2** and **3** proved to possess different HB-donating potential from the titration experiment, the dominant conformations of these compounds were next examined by using calculation techniques (Figs. 2 and 3) [10]. It is reported that aminothiourea **1a** exists as a mixture of several conformers including *anti*- and *syn*-isomers in solution, where the *anti*-form is

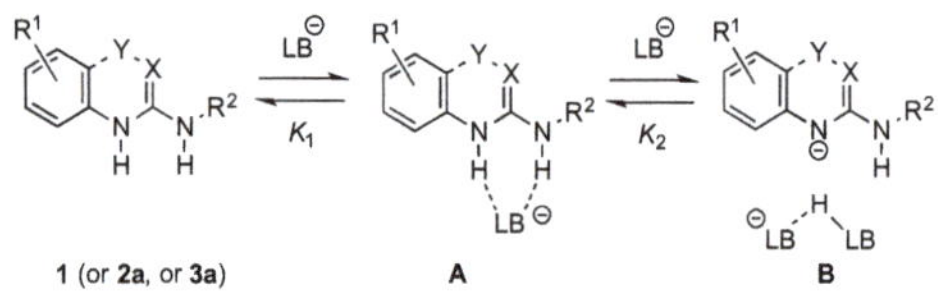


**Scheme 2** Association models in the reaction of the HB donors with Lewis bases (LB): **A** (Cat·LB) and **B** [Cat·(LB)$_2$]

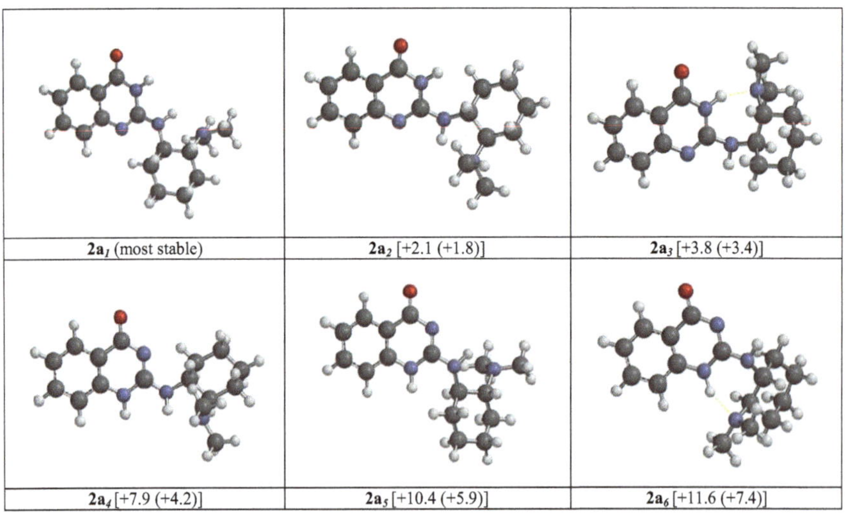


**Fig. 2** The conformers of **2a** identified as energy minima. The relative Gibbs free energies (in kcal/mol) were calculated at B3LYP/6-31G* level using Gaussian 09. PCM-corrected values are shown in *parentheses* (in $CH_2Cl_2$)

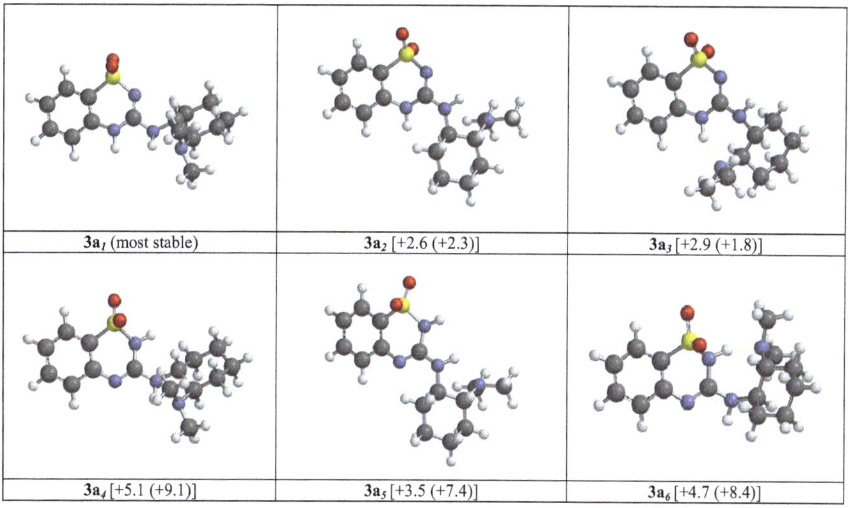


**Fig. 3** The conformers of **3a** identified as energy minima. The relative Gibbs free energies (in kcal/mol) were calculated at B3LYP/6-31G* level using Gaussian 09. PCM-corrected values are shown in *parentheses* (in $CH_2Cl_2$)

rather prevalent [18]. However, the corresponding *syn*-isomers **2a**$_1$ and **3a**$_1$ are suggested to be the energetically most stable conformers in dichloromethane in the cases of quinazoline **2a** and benzothiadiazine **3a**. More interestingly, another *syn*-isomer **2a**$_4$, which adopts the same catalytically active conformation as **1a** and **3a**$_1$, was initially anticipated to be the most stable conformer, but the unexpected *syn*-isomer **2a**$_1$ was shown to be more stable than **2a**$_4$ on the basis of DFT calculations. By taking account of association constants and conformational analyses, benzothiadiazine **3a** would be the most promising HB-donor catalyst.

# 3 Synthetic Applications of Bifunctional Guanidines

## *3.1 Hydrazination*

The asymmetric α-amination of β-keto esters is one of the most important reactions, because this transformation provides optically active natural and unnatural α-amino acid derivatives from readily available starting materials [19]. Since the hydrazination with azodicarboxylates is a reliable method for the α-amination, its asymmetric version has been extensively studied using a variety of chiral catalysts [20–24]. To evaluate the catalytic performance of new bifunctional guanidines, the reaction of keto ester **4a** with di-*tert*-butyl azodicarboxylate was investigated with several HB-donor catalysts (Table 1). The hydrazination of **4a** with 10 mol% of

**Table 1** Hydrazination of **4a** with di-*tert*-butyl azodicarboxylate

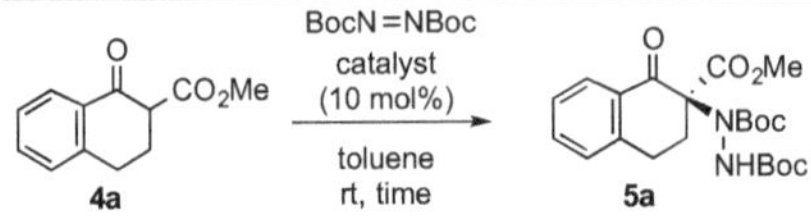


| Entry[a] | Catalyst | Time (h) | Yield (%)[b] | ee (%)[c] |
|---|---|---|---|---|
| 1 | **1a** | 20 | 22 | 83 |
| 2[d] | **1b** | 5 | 99 | 87 |
| 3 | **2a** | 3 | 93 | 96 |
| 4 | **3a** | 3 | 82 | 84 |
| 5 | **2b** | 3 | 93 | 92 |
| 6 | **2c** | 5 | 99 | 95 |
| 7 | **2d** | 3 | 98 | 95 |
| 8 | **2e** | 5 | Quant | 96 |

[a]Unless otherwise noted, the reactions were conducted with **4a** (1.1 equiv.), di-*tert*-butyl diazodicarboxylate (1.0 equiv.), and the catalysts (10 mol%) in toluene at room temperature
[b]Isolated yield
[c]Determined by chiral HPLC analyses
[d]At −40°C

thiourea **1a** in toluene at room temperature gave the desired product **5a** in 22% yield with 83% ee (entry 1). The low yield is attributed to the decomposition of the catalyst by the reaction with the reagent [17]. The use of the corresponding urea catalyst **1b** led to full conversion of **4a**, but the ee was only 87% even at −40°C (entry 2). As expected, the reactions using bifunctional guanidine catalysts **2a** and **3a** proceeded smoothly to afford **5a** in good yields. In particular, the quinazoline-type catalyst **2a**, being the weaker HB-donor catalyst, furnished the product with 96% ee at room temperature (entries 3 and 4), which is a major advantage over the cryogenic conditions required for catalyst **1b**. Further investigations with fluorinated quinazoline derivatives **2b-e** only showed marginal effects on the chemical yield and enantioselectivity (entries 5–8). These results indicated that quinazolines **2a** and **2e** would be the best catalysts for the asymmetric hydrazination.

The 8-fluorinated quinazoline **2e** was applied to the asymmetric hydrazination of 1,3-dicarbonyl compounds and an α-cyanoester as shown in Table 2. Under the optimized conditions, the five- and seven-membered keto esters **6b** and **6c** underwent stereoselective hydrazination, giving satisfactory ee's even at room temperature, while the urea-catalyzed hydrazination of **6b** and **6c** required lowering of the temperature to attain the same enantioselectivities (entries 1 and 2). High selectivity (91% ee) was also obtained from the sterically less-hindered methyl ester **6d** instead of *tert*-butyl esters generally used with **1b** (entry 3). Similarly, more reactive substrates such as 1,3-diketone **6e** and cyanoester **6f** only resulted in moderate ee's (80% and 73% ee) by the treatment with **1b**, but the use of quinazoline **2e** provided the corresponding products **7e** and **7f** in significantly improved enantiomeric excesses (entries 4 and 5).

**Table 2** Scope of substrates **6** for the hydrazination

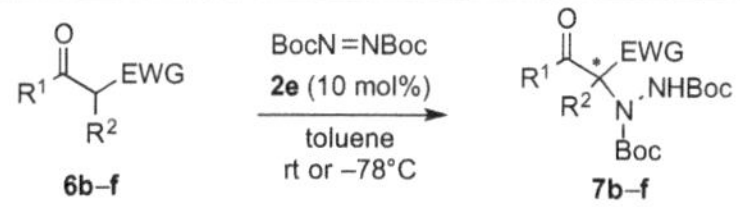


| Entry[a] | Substrate | Temp | Time (h) | Yield (%)[b] | ee (%)[c] |
|---|---|---|---|---|---|
| 1 | 6b 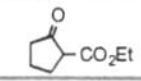 | rt | 4 | 99 | 84 |
| 2 | 6c 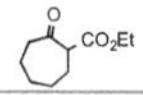 | rt | 78 | 86 | 81 |
| 3 | 6d 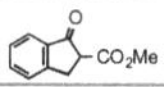 | −78°C | 5 | 94 | 91 |
| 4 | 6e 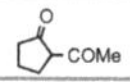 | −78°C | 16 | 87 | 94 |
| 5 | 6f 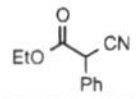 | −78°C | 3 | Quant | 88 |

[a]The reactions were conducted with **6** (1.1 equiv.) di-*tert*-butyl diazodicarboxylate (1.0 equiv.) and the catalysts (10 mol%) in toluene.
[b]Yield of isolated product
[c]Determined by chiral HPLC analyses

Since catalysts **1b** and **2e** afforded the same enantiomers, this reaction was assumed to proceed in a manner similar to that with previously reported HB-donor catalysts [17]. In addition, the superior catalytic performance of **2e** as compared to urea **1b** and benzothiadiazine **3a** might be rationalized by the different tautomeric structure of **2e** compared to those of the latter two, as described in the previous section, as well as the mild association with the substrates. Importantly, these results suggest that the strongest HB donors are not always the best catalysts for asymmetric reactions.

## 3.2 *Isomerization of Alkynoates to Allenoates*

Allenes are versatile and valuable substructures in synthetic organic chemistry due to their characteristic structure and reactivity originated from the contiguous C=C double bonds [25, 26]. In particular, the axial chirality of allene derivatives has attracted much attention from organic chemists from the viewpoint of synthetic utility as chiral synthons. Therefore, there have been so far many reports on the asymmetric synthesis of optically active allenes [27, 28]. Among them, we focused on the base-catalyzed isomerization of alkynylesters into allenoesters from the atom-economical perspective [8, 9, 29–31].

### 3.2.1 Isomerization of α-Unsubstituted 3-Butynoates

The asymmetric isomerization of 3-butynoates **8a** using a chiral guanidine catalyst was reported by the Tan group [31]. The reaction provided the allenoesters in 79–95% ee, but low reaction temperature (−20°C) and high dilution conditions need to be improved. The isomerization of **8a** was examined in toluene with the bifunctional catalysts to screen their catalytic activities (Table 3) [8]. Whereas no reaction occurred without catalysts, the treatment of **8a** with several catalysts **1a**, **2a**, and **3a** furnished the desired product **9a**, respectively, with similar conversions of **8a** to **9a** (entries 1–4). Among the catalysts examined, the benzothiadiazine catalyst **3a** marked the best catalytic activity and highest enantioselectivity. Moreover, the catalytic amount of **3a** could be reduced to 2 mol% in THF, still reaching up to 98% ee (entries 5–8). On the other hand, benzamide **10** gave the product in good ee, albeit with low conversion (entry 9). These results imply that the high recognizing abilities of the HB donors are desirable for the achievement of high enantioselectivity as well as high yield, in contrast to the hydrazination reaction in the previous section. The absolute configuration of **9a** was determined to be (*S*) by the Lowe-Brewster rule [32, 33].

The reaction progress of **8a** was monitored by means of $^1$H NMR and HPLC at regular intervals (Table 4). Surprisingly, the reaction in dichloromethane was saturated within 3 h and the enantioselectivity gradually decreased with time (entry 1). In THF, the reaction proceeded more slowly than in dichloromethane,

**Table 3** Isomerization of **8a** using chiral HB-donor catalysts

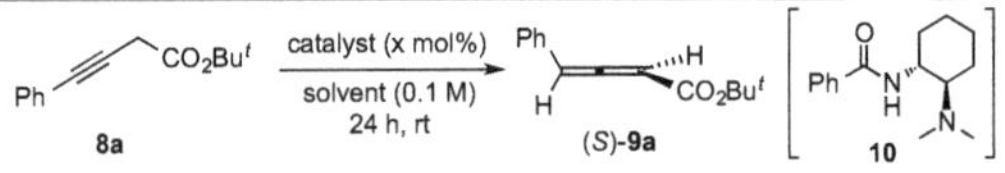


| Entry[a] | Catalyst (*x*) | Solvent | Yield (%)[b] | **8a**:**9a**[c] | ee of **9a** (%)[d] |
|---|---|---|---|---|---|
| 1 | None | Toluene | 100 | 100:0 | – |
| 2 | **1a** (10) | Toluene | 87 | 32:68 | 87 |
| 3 | **2a** (10) | Toluene | 100 | 34:66 | 80 |
| 4 | **3a** (10) | Toluene | 93 | 32:68 | 90 |
| 5 | **3a** (10) | $CH_2Cl_2$ | 100 | 34:66 | 76 |
| 6 | **3a** (10) | THF | 100 | 33:67 | 91 |
| 7 | **3a** (5) | THF | 100 | 34:66 | 96 |
| 8 | **3a** (2) | THF | 98 | 36:64 | 98 |
| 9 | **10** (2) | THF | 100 | 86:14 | 90 |

[a]The reactions were conducted with **8a** (1.0 equiv.) and catalyst (*x* mol%) in several solvents at room temperature
[b]Combined yield of **8a** and **9a** after purification by column chromatography
[c]Determined by $^1$H NMR spectroscopy
[d]Determined by chiral HPLC analyses

**Table 4** Time-course studies of the isomerization of **8a** catalyzed by **3a**

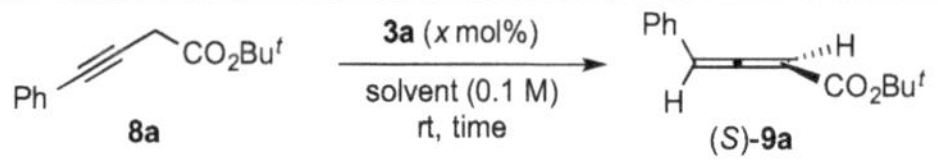


| Entry[a] | *x* | Solvent | 3 h[b] | 6 h[b] | 12 h[b] | 24 h[b] |
|---|---|---|---|---|---|---|
| 1 | 5 | $CH_2Cl_2$ | 37:63 (96) | 36:64 (93) | 36:64 (90) | 34:66 (83) |
| 2 | 5 | THF | 63:37 (94) | 46:54 (96) | 36:64 (95) | 38:62 (92) |
| 3 | 10 | THF | 51:49 (97) | 37:63 (94) | 37:63 (92) | 33:67 (87) |
| 4 | 2 | THF | 82:12 (97) | 68:32 (98) | 51:49 (97) | 39:61 (96) |

[a]The reactions were conducted with **8a** (1.0 equiv.) and **3a** (*x* mol%) at room temperature
[b]The values show the ratio of **8a/9a** and ee values of **9a** in parentheses

**Table 5** Scope of substrates for the isomerization of alkynoates **8b-m**

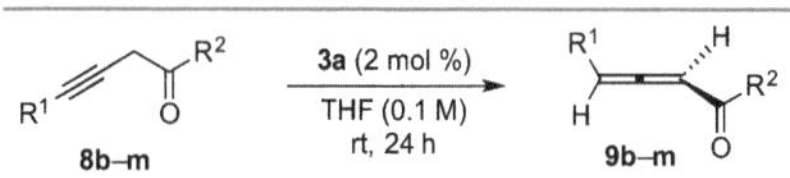


| Entry[a] | 8 ($R^1$, $R^2$) | Yield (%)[b] | **8**:**9**[c] | ee of **9** (%)[d] |
|---|---|---|---|---|
| 1 | **8b** (4-$MeOC_6H_4$, $OBu^t$) | 100 | 47:53 | 97 |
| 2 | **8c** (4-$MeC_6H_4$, $OBu^t$) | 98 | 39:61 | 95 |
| 3 | **8d** (4-$ClC_6H_4$, $OBu^t$) | 100 | 30:70 | 95 |
| 4 | **8e** (2-$MeOC_6H_4$, $OBu^t$) | 91 | 23:77 | 96 |
| 5[e] | **8f** (2-$ClC_6H_4$, $OBu^t$) | 100 | 49:51 | 91 |
| 6 | **8g** (3-$MeC_6H_4$, $OBu^t$) | 100 | 36:64 | 98 |
| 7 | **8h** (2-(CHO)$C_6H_4CH_2$, $OBu^t$) | 96 | 77:23 | 95 |
| 8 | **8i** (BnO$(CH_3)_2$C, $OBu^t$) | 100 | 40:60 | 98 |
| 9 | **8j** (HO$(CH_3)_2$C, $OBu^t$) | 99 | 23:77 | 97 |
| 10 | **8k** (Ph, OEt) | 96 | 33:67 | 97 |
| 11 | **8l** (Ph, $NMe_2$) | 100 | 38:62 | 96 |
| 12 | **8m** (Ph, $N(CH_2)_4$) | 100 | 38:62 | 94 |

[a]Unless otherwise noted, the reactions were conducted with **8** (1.0 equiv.) and **3a** (2 mol%) in THF at room temperature for 24 h
[b]Combined yield of **8** and **9** after purification with column chromatography
[c]Determined by $^1H$ NMR spectroscopy
[d]Determined by chiral HPLC analyses
[e]The reaction was performed for 12 h

and a slight decrease in stereoselectivity was still observed depending on the catalyst loading in this time-course study (entries 2–4). Overall, the reaction should be performed with less than 2 mol% of the catalyst and for just enough time for the reaction to complete.

The scope of substrates isomerized under the optimized conditions (2 mol% of **3a** at room temperature for 24 h) is shown in Table 5. Substrates **8b-d** with electron-donating or electron-withdrawing groups at the *para*-position of the phenyl

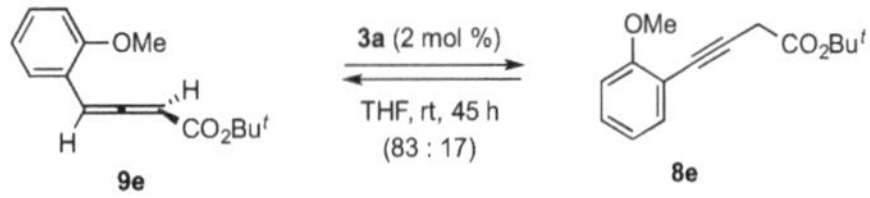


**Scheme 3** Thermodynamic study of the isomerization of allenoate **9e**

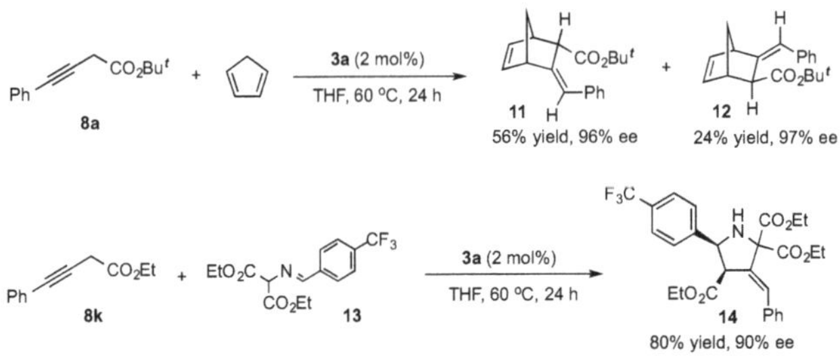


**Scheme 4** Tandem isomerization and cycloaddition of alkynoates **8** catalyzed by **3a**

rings could be converted into the corresponding allenoates **9b-d** with high enantioselectivities (entries 1–3). The position of the substituent on the phenyl ring did not affect the stereoselectivity of the isomerization, and all of the substrates **8e**-**g** achieved good to high ee's (entries 4–6). Similar to the aromatic substrates, alkyl-substituted alkynoates **8h**-**j** underwent the desired reaction smoothly with excellent selectivity using only 2 mol% of the catalyst (entries 7–9). The less-bulky ethyl ester **8k** was tolerated just like *tert*-butyl esters, the excellent ee being fully maintained (entry 10). Although the reactions of alkynyl amides **8l** and **8m** gave somewhat lower conversion compared to those of alkynyl esters, the stereoselectivity remained high (entries 11 and 12).

When the isolated allene **9e** was treated with catalyst **3a** for 45 h, the corresponding alkynoate **8e** was generated as a minor product, with most of **9e** being recovered unchanged (Scheme 3). This result confirms that the isomerization of **8** into **9** in the presence of **3a** is reversible. To complete the conversion of alkynoate into allenoates, the tandem isomerization of alkynoates **8** and cycloaddition of the in situ generated allenes **9** with a proper diene or 1,3-dipole was attempted (Scheme 4) [34]. As expected, when cyclopentadiene was present during the isomerization of **8a**, the subsequent Diels-Alder reaction of the resulting allene and cyclopentadiene occurred [35] to give the corresponding *endo*-cycloadducts **11** in good yield and high ee, even at elevated temperature, together with the minor *exo*-diastereomer **12**. Furthermore, azomethine ylide proved to be useful for the tandem reaction. Using this method, the synthetically useful 3-alkylidene pyrrolidine **14** could be prepared as a single product in a one-pot process from the corresponding alkynoate **8k** and diester **13** without a significant loss of enantioselectivity [36].

### 3.2.2 Isomerization of α-Substituted 3-Butynoates via Dynamic Kinetic Resolution

When α-substituted 3-butynoates **15** [37] are similarly treated with the bifunctional catalysts, the corresponding trisubstituted allenoesters **16** would be obtained via 1,3-proton migration [9]. However, such an asymmetric reaction has not been reported yet. Distinct from achiral α-unsubstituted 3-butynoates, α-substituted 3-butynoates exist as a racemic mixture. Therefore, the catalyst needs to differentiate two enantiomers of racemic alkynyl esters in a highly stereoselective manner for the kinetic resolution. Furthermore, if the same catalyst promotes the racemization of both enantiomers, the dynamic kinetic resolution (DKR) [38–40] of the racemic alkynes would be realized to provide chiral trisubstituted allenoesters in good yield without recovering the other enantiomers, which are less reactive in the 1,3-proton migration (Scheme 5).

Several bifunctional catalysts **1–3** were screened for the isomerization of alkyne **15a**. As shown in Table 6, the ratio of **16a/15a** as well as the enantioselectivity of **16a** were significantly affected by both catalyst and solvent (entries 1–6). Moreover, racemization did in fact occur in the presence of the catalysts to give the

**Scheme 5** Isomerization of α-substituted alkynoate via a (dynamic) kinetic resolution

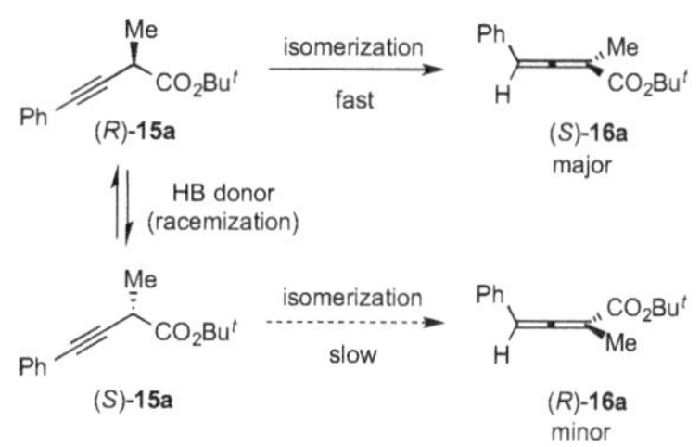


**Table 6** Screening of the reaction conditions

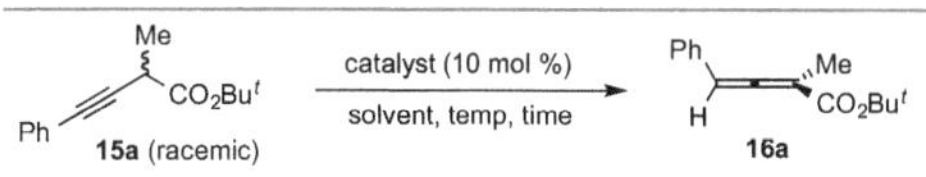


| Entry[a] | Catalyst | Solvent | Temp | Time (h) | Yield (%)[b] | **15a**:**16a**[c] | ee of **16a** (%)[d] |
|---|---|---|---|---|---|---|---|
| 1 | **1a** | $CH_2Cl_2$ | rt | 168 | 89 | 5:95 | 95 |
| 2 | **2a** | $CH_2Cl_2$ | rt | 168 | 85 | 30:70 | 70 |
| 3 | **3a** | $CH_2Cl_2$ | rt | 168 | 91 | 8:92 | 96 |
| 4 | **3a** | Toluene | rt | 168 | 90 | 45:55 | 88 |
| 5 | **3a** | THF | rt | 168 | 94 | 46:54 | 86 |
| 6 | **3a** | $(ClCH_2)_2$ | rt | 168 | 94 | 10:90 | 93 |
| 7 | **3a** | $(ClCH_2)_2$ | 60°C | 24 | 93 | 5:95 | 93 |

[a]Unless otherwise noted, the reactions were conducted with **15a** (1.0 equiv.) and catalyst (10 mol %) in several solvents (0.1 M)
[b]Combined yield of **15a** and **16a** after purification by column chromatography
[c]Determined by $^1$H NMR spectroscopy
[d]Determined by chiral HPLC analyses

desired product **16a** in more than 50% yield. Although both thiourea **1a** and benzothiadiazine **3a** showed good results in terms of yield and ee, the latter was preferentially used due to the enhanced catalytic performance. The reaction time could be shortened by raising the reaction temperature without lowering the enantioselectivity (entry 7).

To gain a deeper insight into the reaction process, the ee's of **15a** and **16a** were traced over time under the optimized conditions (Table 7). The ee's of the substrate **15a** were gradually improved within 24 h from the beginning of the reaction (12–48% ee), which clearly indicates that the kinetic resolution of the racemic substrate **15a** proceeded (entries 1–4). However, the result of entry 4, where almost half of the starting material was consumed, illustrates that the racemization of **15a** concurrently occurs along with the enantioselective isomerization. More importantly, the further extension of reaction times resulted in higher yield and ee of the desired allenoester **16a** (entries 5 and 6). From these outcomes, it would be reasonable to consider that the isomerization from alkyne **15a** to trisubstituted allene **16a** is not a static reaction, distinct from that of α-unsubstituted derivative **8a**, in which the equilibrium between alkyne **8** and disubstituted allene **9** exists. Indeed, the same treatment of the racemic allene **16a** under the optimized conditions with **3a** only recovered the racemic compound **16a** in 88% yield, and none of the corresponding alkyne **15a** was observed in the reaction mixture. The existence of the methyl group at the α-position completely suppresses the reverse reaction from **16a** to **15a**.

The substrate scope of the reaction performed under the optimized reaction conditions is shown in Table 8. Other than *tert*-butyl esters, less-hindered ethyl and ally esters could also be used to furnish the corresponding allenoesters **16b** and **16c** in good yields, albeit with slightly lower ee's (entries 1 and 2). The alkynes **15d-h** bearing a variety of aryl and heteroaryl groups as the $R^1$ substituent also underwent isomerization in a stereoselective manner (entries 3–7), but the enantiomeric excesses were slightly influenced by the position of the substituents of the aryl group (82–94% ee's). Similarly, the subjection of alkynyl ester **15i** bearing an ethyl group as the $R^2$ substituent to the same conditions provided the allenoester **16i**

**Table 7** Time course of the isomerization

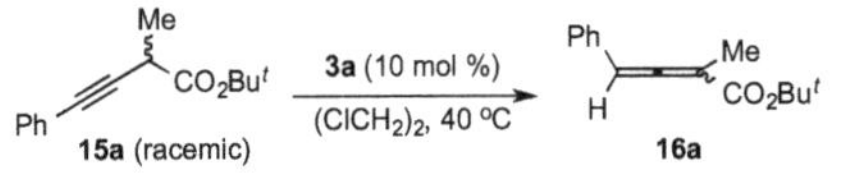


| Entry[a] | Time (h) | **15a**:**16a**[b] | ee of **15a** (%)[c] | ee of **16a** (%)[c] |
|---|---|---|---|---|
| 1 | 2 | 86:14 | 12 | 88 |
| 2 | 6 | 76:24 | 24 | 90 |
| 3 | 12 | 56:44 | 42 | 90 |
| 4 | 24 | 33:67 | 48 | 91 |
| 5 | 36 | 26:74 | 44 | 94 |
| 6 | 108 | 7:93 | 34 | 94 |

[a]The reactions were conducted with **15a** (0.3 mmol) and **3a** (10 mol%) in $ClCH_2CH_2Cl$ at 40°C
[b]Determined by $^1H$ NMR spectroscopy
[c]Determined by chiral HPLC analyses

**Table 8** Scope of substrates **15** catalyzed by **3a**

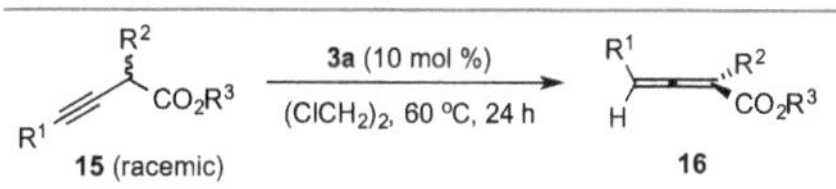


| Entry[a] | **15** ($R^1$, $R^2$, $R^3$) | Yield (%)[b] | **15**:**16**[c] | ee of **16** (%)[d] |
|---|---|---|---|---|
| 1 | **15b** ($C_6H_5$, Me, Et) | 96 | 1:99 | 87 |
| 2 | **15c** ($C_6H_5$, Me, allyl) | 96 | 1:99 | 86 |
| 3 | **15d** (4-MeO$C_6H_4$, Me, Bu$^t$) | 93 | 8:92 | 92 |
| 4 | **15e** (4-Cl$C_6H_4$, Me, Bu$^t$) | 90 | 4:96 | 92 |
| 5 | **15f** (3-Cl$C_6H_4$, Me, Bu$^t$) | 91 | 1:99 | 82 |
| 6 | **15g** (2-Cl$C_6H_4$, Me, Bu$^t$) | 96 | 17:83 | 85 |
| 7 | **15h** (3-thienyl, Me, Bu$^t$) | 96 | 12:88 | 94 |
| 8 | **15i** ($C_6H_5$, Et, Bu$^t$) | 96 | 4:96 | 87 |

[a]Unless otherwise noted, the reactions were conducted with **15** (1.0 equiv.) and **3a** (10 mol%) in $ClCH_2CH_2Cl$ (0.1 M)
[b]Combined yield of **15** and **16** after purification by column chromatography
[c]Determined by $^1H$ NMR spectroscopy
[d]Determined by chiral HPLC analyses

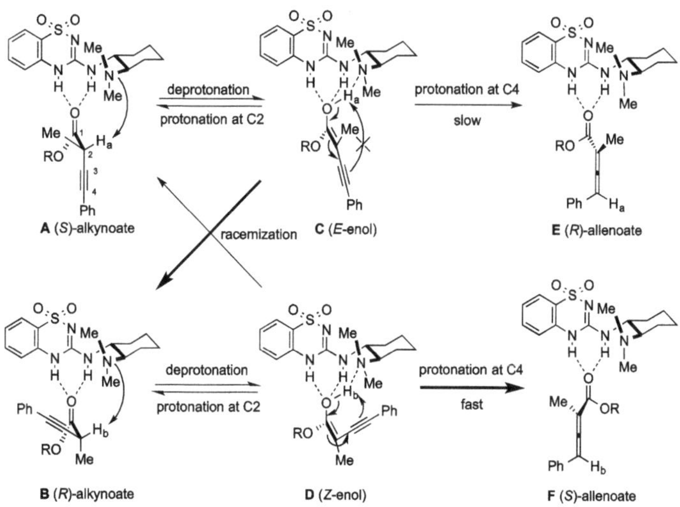


**Scheme 6** Plausible mechanism of the **3a**-catalyzed isomerization of **15a**. **A**: the binary complex of **3a** and (*S*)-**15**, **B**: the binary complex of **3a** and (*R*)-**15**, **C**: the intermediate *E*-enol activated by **3a**, **D**: the intermediate *Z*-enol activated by **3a**, **E**: a binary complex of **3a** and (*R*)-allenoate, **F**: the binary complex of **3a** and (*S*)-allenoate

with 87% ee (entry 8). In the syntheses of trisubstituted allenes **16a-i**, the good to high conversion of the starting materials was consistently observed irrespective of the $R^1$ and $R^2$ substituents. The absolute configuration of the obtained products **16a-i** was again determined to be *S* according to the Lowe-Brewster's rule [32, 33].

Based on the absolute configuration of the products, a plausible reaction mechanism [41] was proposed as shown in Scheme 6. The (*E*)-enol complex **C** would be generated stereoselectively via the deprotonation of the α-proton ($H_a$) from the binary complex **A** of (*S*)-alkynoate **15** and catalyst **3a**. In the complex **A**, the two N-H protons of the catalyst activate the carbonyl group of the ester through the double hydrogen bonding interaction to enhance the acidity of the α-proton ($H_a$). Consequently, the $H_a$ proton gets closer to the tertiary amine of the catalyst, and the following deprotonation proceeds much faster than the uncatalyzed reaction. Similarly, the reaction of (*R*)-alkynoate **15** with **3a** provides the (*Z*)-enol complex **D** by the stereoselective deprotonation of $H_b$ in the complex **B**. The resulting ammonium proton ($H_b$) of the complex **D** could migrate to the C2 and C4 positions, giving the original complex **B** and the desired (*S*)-allenoate complex **F**, respectively. The latter process is considered to proceed efficiently via the energetically more stable six-membered transition state, yielding the (*S*)-allenoate predominantly. On the other hand, in the case of complex **C**, the same migration of $H_a$ to the C4 position would be difficult due to the trans geometry of $H_a$ and the alkynyl group, leading to the diastereomeric complexes **A** and **B** via the racemization of (*S*)-alkynoate. Overall, the major product, (*S*)-allenoate, would be formed through the 1,3-proton migration of (*R*)-alkynoate, which is alternatively supplied from (*S*)-alkynoate via the non-stereoselective protonation at the C2 position of the (*E*)-enol complex **C**.

### *3.3 The Intramolecular Oxa-Michael Addition to α,β-Unsaturated Amides and Esters*

Chiral oxygen-containing heterocycles, such as isoxazolines, isoxazolidines [42, 43], chromans [44], and dihydrobenzofurans [45], can be found in numerous natural products and biologically active compounds. In particular, the representative *O*-heterocycles bearing an acetamide or acetate as a side chain are depicted in Fig. 4. Significant effort has been directed toward the development of synthetic methods for accessing these materials in an atom- and step-economic manner. One of the most promising approaches to supply these chiral *O*-heterocycles involves the asymmetric intramolecular oxa-Michael (AIOM) reaction [46–49] of α,-β-unsaturated carboxylic acid derivatives such as thioesters [50] and imides [51], while several additional steps are required for the conversion of the obtained products into the target molecules as well as the activation of carboxylic acids to

**Fig. 4** *O*-Heterocycles bearing an acetamide or acetate

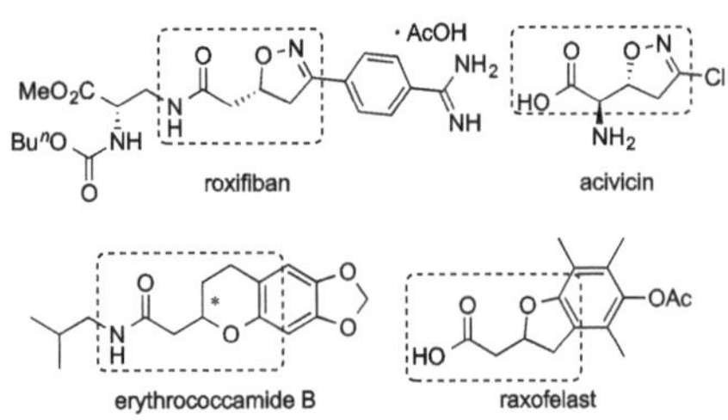

the corresponding thioesters or imides. Moreover, in the oxa-Michael adducts of a thioester or imide, there is the risk of racemization via retro-Michael/Michael addition during these transformations. Therefore, the direct AIOM addition to unactivated α,β-unsaturated esters [52–56] or amides [11] would be more desirable from the perspective of substrate availability and tolerance to racemization of products. However, such a reaction has received much less attention, likely because of the poor reactivity of both the Michael acceptor and the *O*-nucleophiles employed.

### 3.3.1 The Intramolecular Oxa-Michael Addition of Hydroxylamine Derivatives

The α,β-unsaturated amide **17a**, which bears a hydroxylamine moiety, was selected as a substrate for the screening of catalysts for the AIOM reaction (Table 9). The *N*-Cbz-hydroxylamine is not only a strong nucleophile but also an active leaving group. Therefore, the use of strong bases for the activation of the nucleophiles might result in low enantioselectivity due to racemization via retro-Michael addition. It is envisaged that, instead of increasing the basicity of the catalysts, improving their HB-donating ability could activate the poorly reactive Michael acceptor to accelerate the cyclization efficiently. Indeed, although general organic bases such as triethylamine and cinchonidine **19** were not effective in this reaction, bifunctional thiourea [16] **1a** and squaramide [57] **20** moderately enhanced the cyclization of **17a** to afford isoxazolidine **18a** in acceptable enantioselectivities (entries 1–3),

**Table 9** Screening of the reaction conditions

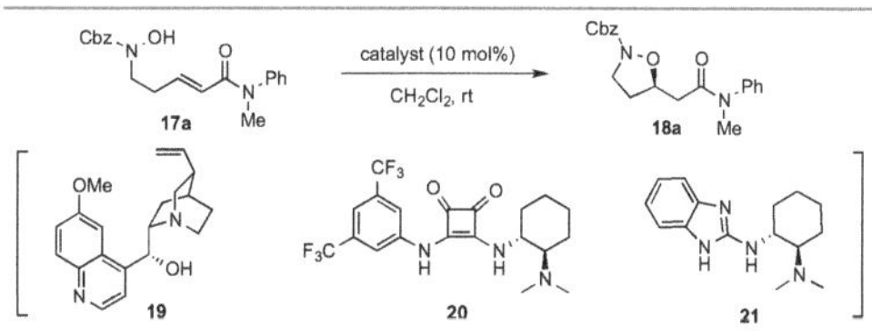


| Entry | Catalyst | Time (h) | Yield[a] (%) | ee[b] (%) |
|---|---|---|---|---|
| 1 | **1a** | 120 | 88 | 81 |
| 2 | **19** | 24 | Trace | n.d. |
| 3 | **20** | 96 | 88 | 88 |
| 4 | **21** | 24 | Trace | n.d. |
| 5 | **2a** | 336 | 74 | 51 |
| 6 | **3a** | 24 | 92 | 95 |
| 7 | **3b** | 24 | 99 | 93 |
| 8 | **3c** | 24 | 99 | 97 |
| 9 | **3d** | 24 | 99 | 79 |

[a]Yields of isolated products
[b]The ee values were determined by chiral HPLC analysis
*n.d.* not determined

albeit long reaction times (4 or 5 days) were needed to complete the reaction. In contrast to these results, the new guanidine-type catalysts **2**, **3**, and **21** displayed distinct catalytic activities in the cyclization, that is, both the chemical yield and enantioselectivity were influenced in relation to the HB-donating ability (entries 4–6). In sharp contrast to benzimidazole [58, 59] **21** and quinazoline [8] **2a**, the reaction with benzothiadiazine [8] **3a** was complete within 24 h, giving the desired product **18a** in 92% yield with 95% ee. These results suggested that the HB-donating strength of the catalyst plays a crucial role in the reaction by tightly binding to amide **17a**. Further experiments with a series of the fluorine-containing benzothiadiazine catalysts **3b-d** revealed that **3c** gave the best enantioselectivity (entries 7–9), presumably because of the catalytic activity derived from the balance between the inductive and mesomeric effects of the fluorine substituent [60].

The catalyst **3a** could be applied to a variety of tertiary and secondary amides **17b-f**, and the desired cyclized adducts **18b-f** were obtained in the range of 70–99% yield (Table 10). The enantioselectivities were somewhat affected by the substituents ($R^2$ and $R^3$) of the amide moiety, more than 90% ee being achieved in the cases of *N*-methylanilides and benzylamide. In addition to the biological importance of these compounds, isoxazolidines **18** are regarded as β-hydroxy-δ-amino acid equivalents, following the cleavage of the N–O bond. To highlight the synthetic utility of **18**, a concise formal total synthesis of atorvastatin [61] was performed from the Michael adduct **18e** (Scheme 7). Thus, amide **18e** was converted into a known synthetic intermediate **22** of atorvastatin by the three-step sequence consisting of esterification, N–O bond cleavage, and cross-Claisen condensation with *tert*-butyl acetate.

**Table 10** Synthesis of chiral isoxazolidines **18b-f**

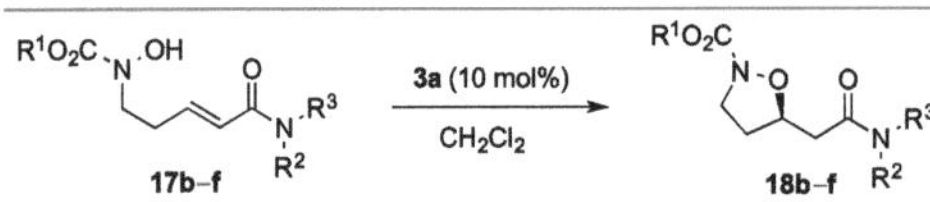


| Entry | Product | Conditions | Yield (%)[a] | ee (%)[b] |
|---|---|---|---|---|
| 1 | 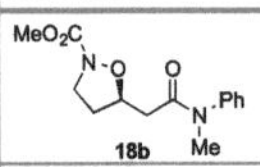 | rt, 24 h | 89 | 92 |
| 2 | 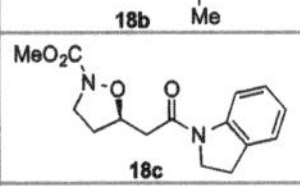 | rt, 24 h | 97 | 84 |
| 3 | 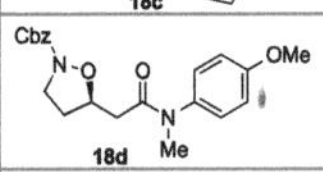 | rt, 24 h | 99 | 96 |
| 4 | 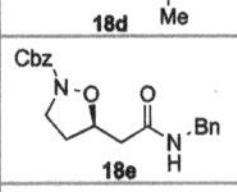 | 40°C, 24 h | 99 | 90 |
| 5 | 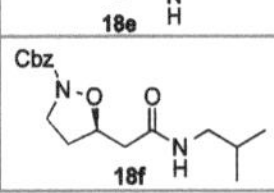 | 40°C, 72 h | 70 | 84 |

[a]Isolated yield
[b]Determined by chiral HPLC analysis

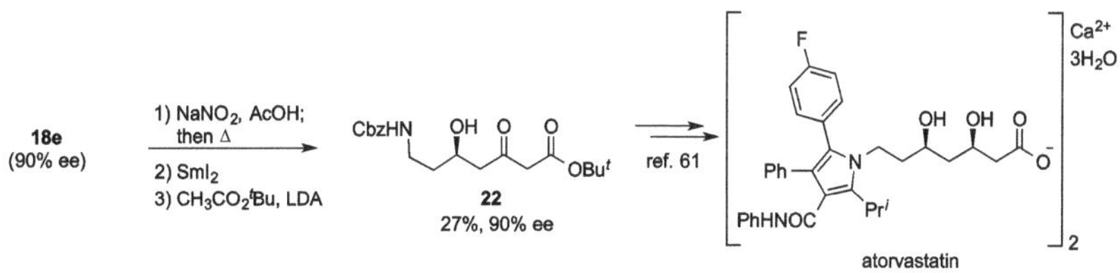


**Scheme 7** A formal synthesis of atorvastatin

### 3.3.2 The Intramolecular Oxa-Michael Addition of Phenols

Since various types of *O*-heterocycles such as chromans and dihydrobenzofurans as shown in Fig. 4 prevail in natural products and pharmaceuticals, the AIOM reaction of α,β-unsaturated amides and esters bearing phenolic OH groups would be useful and practical from the viewpoint of atom economy. So far, there have been only a few reports on the same reaction of α,β-unsaturated esters [52–56], and no reports with regard to α,β-unsaturated amides have yet been described. Recently we discovered that benzothiadiazine catalyst **3a** effectively promoted the AIOM addition to α,β-unsaturated amides **23a-i** (Table 11) [11]. A variety of dihydrobenzofurans **24a-g** and chromans **24h-i** were synthesized in good yields (68–99%) and high enantioselectivities (83–98% ee), with methyl, methoxy, nitro, and bromo groups being well tolerated under the reaction conditions. A gram-scale reaction of **23b** (1.15 g, 5.2 mmol) cleanly furnished the amide **24b** (1.12 g, 97%, 92% ee) with only 1.0 mol% of the catalyst **3c**, albeit with a longer reaction time. In sharp contrast with benzothiadiazine **3a**, the same reactions of **23a**, **23d**, **23h**, and **23i** with the thiourea **1a** clearly demonstrate the superior catalytic potential of the benzothiadiazine catalysts **3a** and **3c**. The chemical yields and ee's were improved considerably by using the benzothiadiazine catalysts **3a** and **3c**.

In addition, the current method could be applied to the AIOM addition of the ester congeners **25a-d** to provide the corresponding dihydrobenzofurans **26a-d** with higher enantioselectivities (86–89% ee) as compared to literature precedents [52–56]. It is worthy to note that the construction of a chiral tetrasubstituted carbon center was also successful, and the corresponding *O*-heterocycle **26d** was obtained in 75% yield with 86% ee. More importantly, an aliphatic OH group could be used as the nucleophile in the cyclization of **25e** to furnish the chiral tetrahydrofuran **26e** with good enantioselectivity.

The advantage of using the current AIOM reaction of phenolic α,β-unsaturated amides is best demonstrated by the rapid asymmetric total synthesis of the natural product erythrococcamide B [62] (Scheme 8). The readily available lactone **27** was subjected to DIBAL and then a HWE reaction to give α,β-unsaturated amide **28**. The first enantioselective synthesis of erythrococcamide B was subsequently accomplished using the **3c**-catalyzed oxa-Michael addition in 59% overall yield with 82% ee. An alternative example of the high catalytic activity of **3c** was further displayed by the first enantioselective synthesis of raxofelast [63, 64], which is used as an antioxidant to modulate inflammatory response. The AIOM reaction of the

**Table 11** Scope of substrates for the oxa-Michael addition of amides **23a-i** and esters **25a-e**

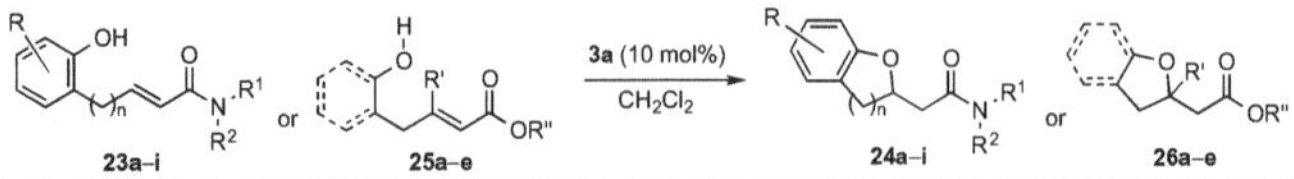


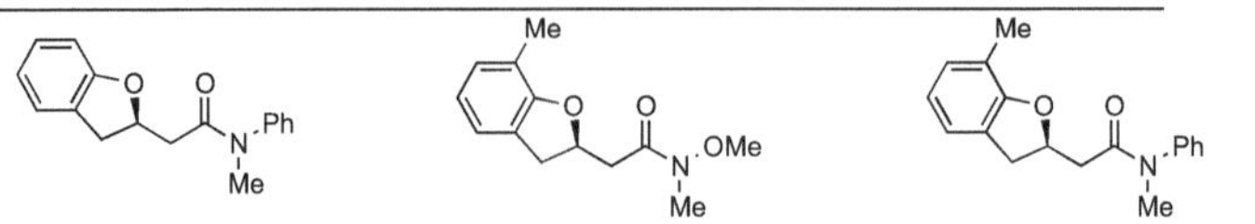

**24a**: 97%, 92% ee (50 °C, 96 h)
with **1a** (10 mol%):
64%, 83% ee (50 °C, 336 h)

**24b**: 70%, 90% ee (rt, 24 h)
with **3c** (1 mol%):
97%, 92% ee (rt, 168 h)

**24c**: 76%, 98% ee (rt, 24 h)

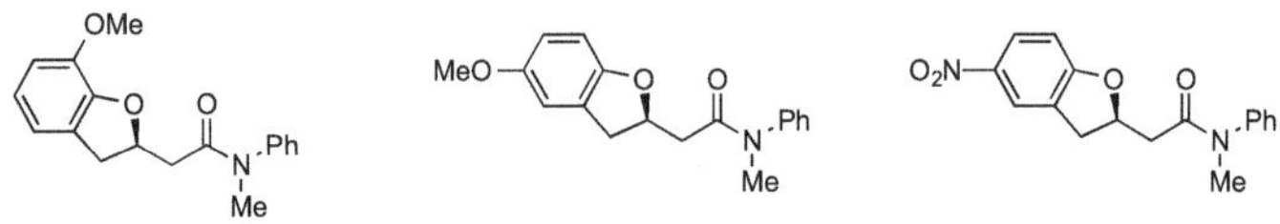

**24d**: 99%, 87% ee (50 °C, 336 h)
with **1a** (10 mol%):
47%, 68% ee (50 °C, 336 h)

**24e**: 83%, 96% ee (rt, 96 h)

**24f**: 76%, 86% ee (rt, 96 h)

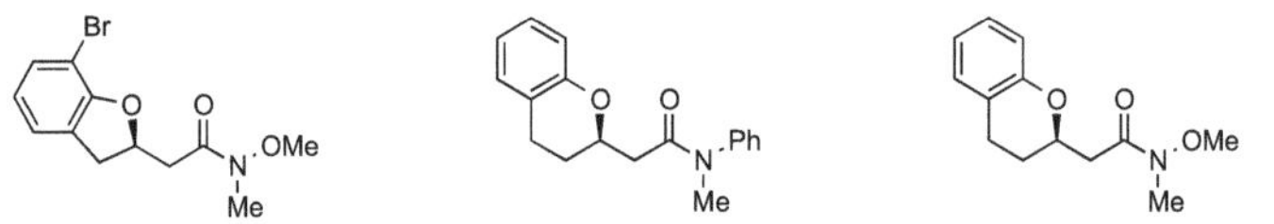

**24g**: with **3c** (1 mol%)
84%, 83% ee (rt, 96 h)

**24h**: 68%, 96% ee (rt, 24 h)
with **1a** (10 mol%):
38%, 79% ee (rt, 24 h)

**24i**: 88%, 89% ee (rt, 24 h)
with **1a** (10 mol%):
74%, 71% ee (rt, 24 h)

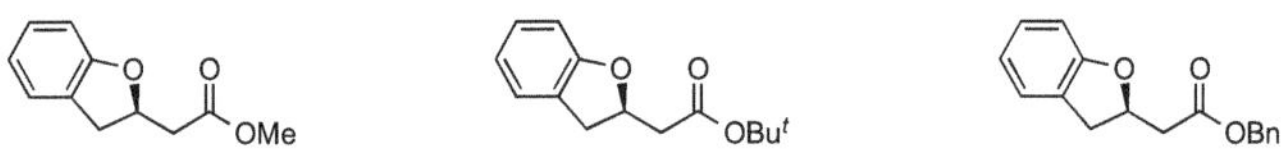

**26a**: 98%, 89% ee (rt, 24 h)

**26b**: 92%, 88% ee (rt, 48 h)

**26c**: 88%, 89% ee (rt, 72 h)

**26d**: 75%, 86% ee (rt, 168 h)

**26e**: 41%, 84% ee (rt, 168 h)

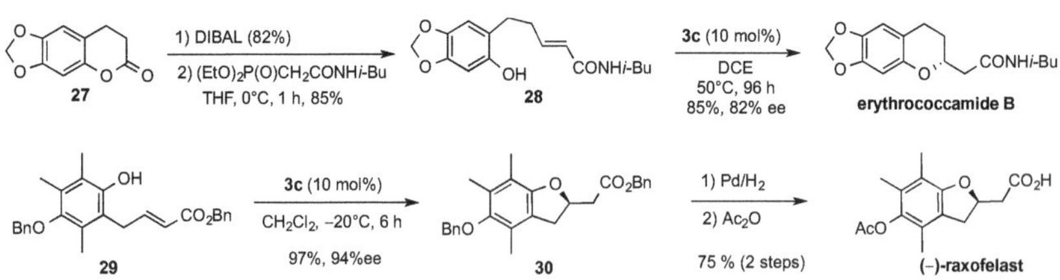


**Scheme 8** Asymmetric synthesis of biologically active natural products

α,β-unsaturated ester **29** not only proceeded at an unprecedentedly low temperature (−20°C) but also afforded the desired product **30**, which was followed by the double debenzylation and the selective acetylation of the resulting phenolic OH group to give the target molecule in 73% overall yield with 94% ee.

## 3.4 The Epoxidation of α,β-Unsaturated Amides and Esters

Chiral epoxides have attracted much attention from synthetic chemists in terms of their importance and versatility as building blocks [65]. One of the reasons would be that the ring opening of these epoxides with a wide range of nucleophiles generally proceeds in a highly or often completely stereo- and/or regioselective manner to afford two continuous stereogenic centers at the same time [66]. Therefore, extensive studies have been done for the synthesis of chiral epoxides, and enormous progress has been made in the synthetic development in the last few decades. Compared with asymmetric electrophilic epoxidations including the Sharpless-Katsuki epoxidation [67] of electron-rich alkenes, asymmetric nucleophilic epoxidation of electron-deficient alkenes has been less explored [68, 69], despite the importance of the produced epoxides. Among them, there has been much interest in 2-oxiranecarboxylic acids and 2-oxiranecarboxamides because of their high potential for further derivatization and their prevalence in several naturally occurring products such as cerulenin, fusarin C, and cyclopenin, which have interesting biological activities (Fig. 5).

In contrast to the epoxidation of enals and enones with chiral primary and secondary amines, only a few successful examples of that of α,β-unsaturated amides have been reported to date [70–73], probably because of the inherent tendencies of the organocatalysts, particularly thioureas, to be oxidized in the presence of an oxidant as well as the relatively poor reactivity of the amide Michael acceptor. Distinct from the thiourea catalyst, bifunctional benzothiadiazine catalysts [8], which have no reactive sulfur atoms and strong hydrogen-bond donating abilities, are expected to efficiently accelerate the asymmetric epoxidation of unreactive α,β-unsaturated amides. When the reaction of acrylamide **31a** and TBHP was carried out in $CH_2Cl_2$ with several catalysts, as shown in Table 12, benzothiadiazine **3a** was shown to be the best catalyst, providing the epoxide **32a** in 90% with the highest enantioselectivity of 70% ee (entries 1–6) [12]. Indeed, the use of other catalysts resulted in incomplete conversion and moderate ee's. Under

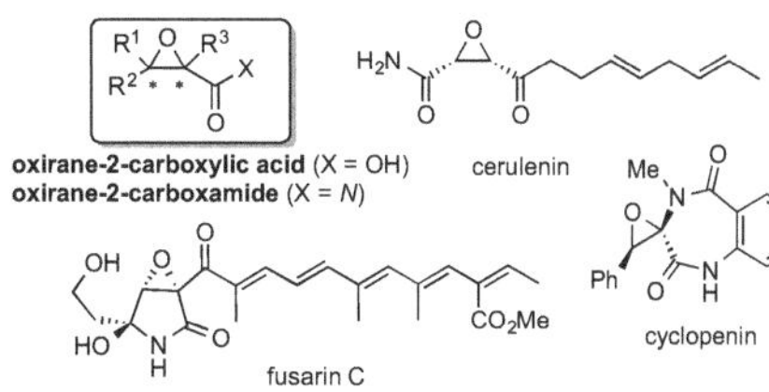


**Fig. 5** Naturally occurring oxiranecarboxylic acid surrogates

**Table 12** Optimization of reaction conditions

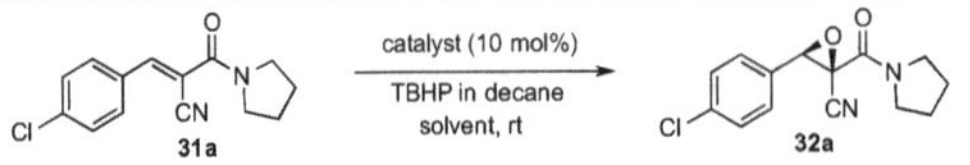

| Entry | Catalyst | Solvent | Time (h) | Yield[a] (%) | ee[b] (%) |
|---|---|---|---|---|---|
| 1 | **1a** | $CH_2Cl_2$ | 24 | 0 | n.d. |
| 2 | **1b** | $CH_2Cl_2$ | 24 | 43 | 45 |
| 3 | **20** | $CH_2Cl_2$ | 24 | 24 | 40 |
| 4 | **21** | $CH_2Cl_2$ | 24 | 49 | 54 |
| 5 | **2a** | $CH_2Cl_2$ | 24 | 27 | 44 |
| 6 | **3a** | $CH_2Cl_2$ | 4 | 90 | 70 |
| 7[c] | **3a** | $CH_2Cl_2$ | 21 | 94 | 20 |
| 8[d] | **3a** | $CH_2Cl_2$ | 24 | 76 | 24 |
| 9 | **3a** | Toluene | 2 | 92 | 57 |
| 10 | **3a** | THF | 72 | 60[e] | 55 |
| 11 | **3a** | MeCN | 72 | 82 | 70 |
| 12 | **3e** | MeCN | 24 | 98 | 84 |

[a]Isolated yield
[b]The ee values were determined by chiral HPLC analysis
[c]30% $H_2O_2$ was used instead of TBHP
[d]Urea hydrogen peroxide was used instead of TBHP
[e]Obtained as an inseparable mixture of diastereomers
*n.d.* not determined

the optimized conditions using the modified catalyst **3e** with the bulky substituent and switching the solvent from $CH_2Cl_2$ to MeCN, the ee value was improved to 84% (entry 12).

The substrate scope, in which a series of different substituents $R^1$ and $R^2$ on the amides were examined, is depicted in Table 13. When *N,N*-dimethylacrylamide was used as the substrate, the corresponding epoxide **32b** was obtained in 89% yield and 74% ee (entry 1), but the enantiomeric excesses obtained with more electron-deficient amides such as morpholinylamide **32c**, anilides **32d** and **32e**, and Weinreb amides **32f** and **32g** were lower at 42–60% ee (entries 2–6). However, it should be noted that almost enantiomerically pure anilides **32d** and **32e** can be obtained after a single recrystallization from ethanol (entries 3 and 4). The reaction of amides **31h** and **31i**, bearing a removable benzyl group, resulted in moderate enantioselectivities (entries 7 and 8), albeit in excellent yields. In sharp contrast to these results, the epoxidation of pyrrolidinylamides **31j-o** bearing different $R^3$ groups as the β-substituent proceeded smoothly to furnish the desired products **32j-o** in 92–98% yield and 80–84% ee (entries 9–14), although the reaction took longer to complete in some cases (entries 12 and 13) due to the electron-donating and steric effect.

Finally, the great usefulness of this type of epoxides **32** was further illustrated by their transformation into a biologically important scaffold [74] (Scheme 9). The

**Table 13** Scope of substrates **31b-p** for the epoxidation

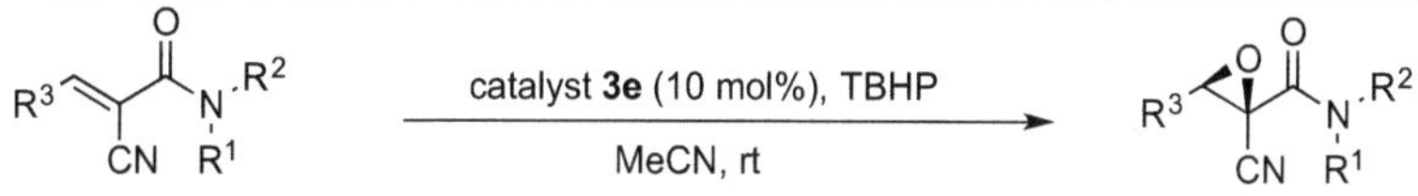


**31b–p** → **32b–p**

| Entry | Product | Time (h) | Yield[a] (%) | ee[b] (%) |
|---|---|---|---|---|
| 1 | 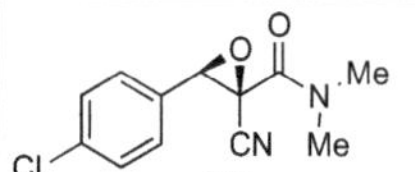 **32b** | 36 | 89 | 74 |
| 2 | 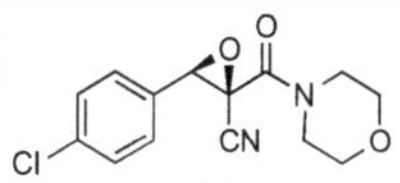 **32c** | 96 | 99 | 60 |
| 3 | 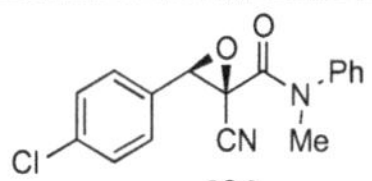 **32d** | 48 | 95 | 52 (99)[c] |
| 4 | 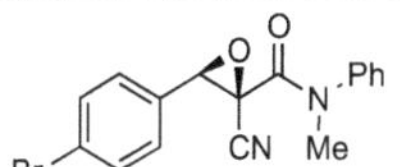 **32e** | 96 | 99 | 52 (95)[c] |
| 5 | 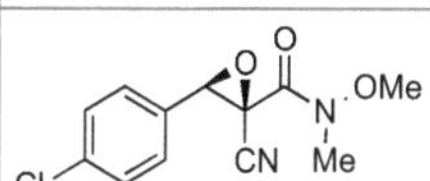 **32f** | 36 | 97 | 46 |
| 6 | 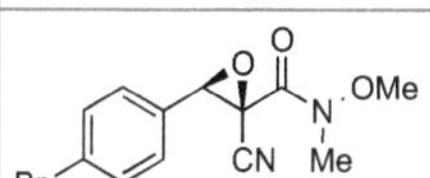 **32g** | 36 | 96 | 42 |
| 7 | 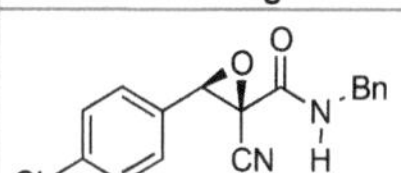 **32h** | 36 | 99 | 53 |
| 8 | 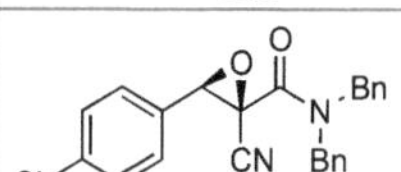 **32i** | 120 | 99 | 33 |
| 9 | 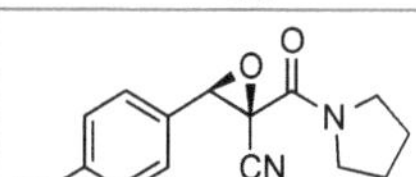 **32j** | 18 | 98 | 84 |
| 10 | 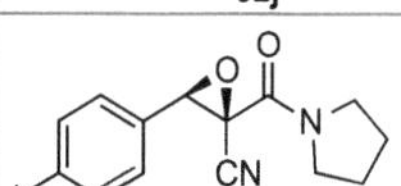 **32k** | 24 | 98 | 84 |
| 11 | 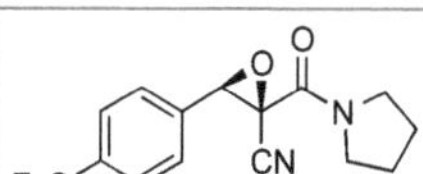 **32l** | 18 | 97 | 80 |

(continued)

Table 13 (continued)

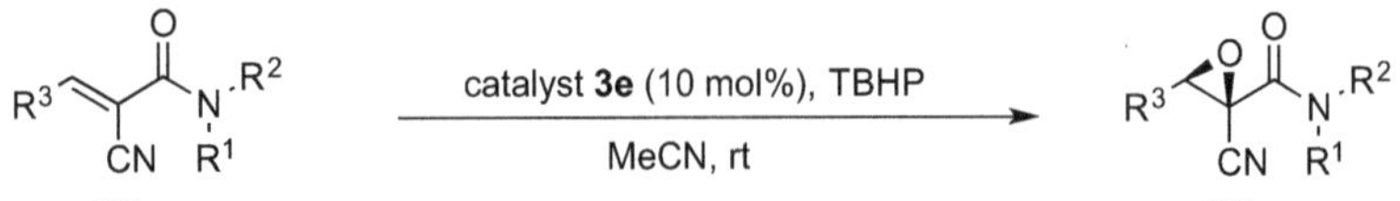


31b–p 32b–p

| Entry | Product | Time (h) | Yield[a] (%) | ee[b] (%) |
|---|---|---|---|---|
| 12 | 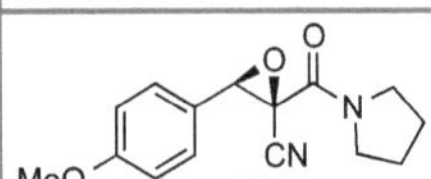 32m | 252 | 96 | 82 |
| 13 | 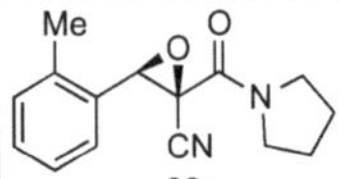 32n | 192 | 92 | 80 |
| 14 | 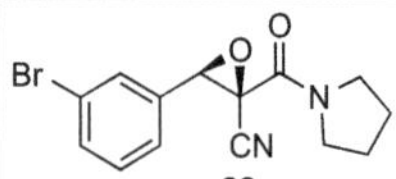 32o | 18 | 97 | 82 |

[a]Isolated yield
[b]Determined by chiral HPLC analysis
[c]The values after single recrystallization from EtOH
*Bn* benzyl

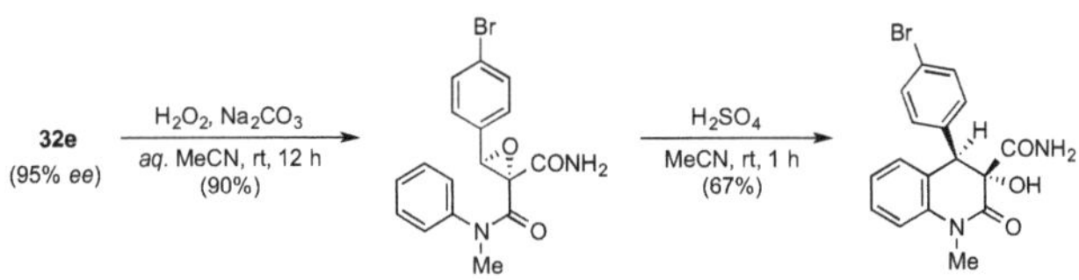


33 34 (96% ee)

**Scheme 9** Synthetic application of epoxide **32e** to chiral 2-quinolone **34**

nitrile group of the epoxide **32e** was chemoselectively hydrolyzed to give the corresponding amide **33** in 90% yield, which was then subjected to sulfuric acid-mediated epoxide-arene cyclization [75], without any loss of enantioselectivity, to give chiral quinolone **34** bearing two contiguous stereogenic centers.

# 4 Conclusion

In this chapter, it is clarified that new guanidine-type hydrogen-bond donors such as quinazoline and benzothiadiazine possess a different mode of substrate recognition as well as distinct hydrogen-bond donating ability as compared to ureas and thioureas. Furthermore, some of these guanidine-type catalysts exhibit excellent catalytic performance in asymmetric reactions, which cannot be achieved with bifunctional thiourea catalysts. The important point is that, since the stronger HB-donor catalyst is not always the best catalyst, universal tuning of hydrogen-bond donor, Brønsted base, and chiral scaffold are required for each reaction. For this purpose, the present lineup of HB donors is not enough, and further explorations to discover innovative HB-donor moieties need to continue for the successful extension of organocatalysis.

The potential of organocatalysts of this particular type would be further broadened by refining the design of the whole structure of catalysts/substrates and identifying applicable reactions on the basis of detailed catalytic mechanisms clarified by computational chemistry.

## References


1. Ishikawa T (2009) Superbases for organic synthesis: guanidines, amidines, phosphazenes and related organocatalysts. Wiley, Hoboken
2. Leow D, Tan C-H (2009) Chem Asian J 4:488
3. Ishikawa T, Kumamoto T (2006) Synthesis 38:737
4. Sohtome Y, Nagasawa K (2010) Synlett 21:1
5. Leow D, Tan C-H (2010) Synlett 21:1589
6. Terada M (2010) J Synth Org Chem Jpn 68:1159
7. Auvil TJ, Schafer AG, Mattson AE (2014) Eur J Org Chem 13:2633
8. Inokuma T, Furukawa M, Uno T, Suzuki Y, Yoshida K, Yano Y, Matsuzaki K, Takemoto Y (2011) Chem Eur J 17:10470
9. Inokuma T, Furukawa M, Suzuki Y, Kimachi T, Kobayashi Y, Takemoto Y (2012) ChemCatChem 4:983
10. Kobayashi Y, Inokuma T, Takemoto Y (2013) J Synth Org Chem Jpn 71:491
11. Kobayashi Y, Taniguchi Y, Hayama N, Inokuma T, Takemoto Y (2013) Angew Chem Int Ed 52:11114
12. Kobayashi Y, Li S, Takemoto Y (2014) Asian J Org Chem 3:403
13. Doyle AG, Jacobsen EN (2007) Chem Rev 107:5713
14. Connon SJ (2008) Chem Commun 44:2499
15. Zhang Z, Schreiner PR (2009) Chem Soc Rev 38:1187
16. Takemoto Y (2010) Chem Pharm Bull 58:593
17. Xu X, Yabuta T, Yuan P, Takemoto Y (2006) Synlett 17:137
18. Hamza A, Schubert G, Soós T, Pápai I (2006) J Am Chem Soc 128:13151
19. Engelbert C (2008) Organic reactions, vol 72. John Wiley and Sons, Inc., Hoboken, pp 1–366
20. Janey JM (2005) Angew Chem Int Ed 44:4292
21. Terada M, Nakano M, Ube H (2006) J Am Chem Soc 128:16044
22. Mashiko T, Hara K, Tanaka D, Fujiwara Y, Kumagai N, Shibasaki M (2007) J Am Chem Soc 129:11342
23. He R, Wang X, Hashimoto T, Maruoka K (2008) Angew Chem Int Ed 47:9466
24. Konishi H, Lam TY, Malerich JP, Rawal VH (2010) Org Lett 12:2028
25. Tius MA (2003) Acc Chem Res 36:284
26. Ma S (2003) Acc Chem Res 36:701
27. Yu S, Ma S (2011) Chem Commun 47:5384
28. Ogasawara K (2009) Tetrahedron Asymmetry 20:259
29. Hashimoto T, Sakata K, Tamakuni F, Dutton MJ, Maruoka K (2013) Nat Chem 5:240
30. Oku M, Arai S, Katayama K, Shioiri T (2000) Synlett 11:493

31. Liu H, Leow D, Huang K-W, Tan C-H (2009) J Am Chem Soc 131:7212
32. Lowe G (1965) Chem Commun 1:411
33. Brewster JH (1967) Top Stereochem 2:1
34. Ma S (2005) Chem Rev 105:2829
35. Jung ME, Nishimura N (2001) Org Lett 3:2113
36. Yu J, He L, Chen X, Song J, Chen W, Gong L (2009) Org Lett 11:4946
37. Hashimoto T, Sakata K, Maruoka K (2009) Angew Chem Int Ed 48:5014
38. Pellissier H (2011) Tetrahedron 67:3769
39. Kim Y, Park J, Kim M-J (2011) ChemCatChem 3:271
40. Lee JH, Han K, Kim M-J, Park J (2010) Eur J Org Chem 2010:999
41. Huang D, Qin S, Hu C (2011) Org Biomol Chem 9:6034
42. Gothelf KV, Jørgensen KA (1998) Chem Rev 98:863
43. Kanemasa S (2002) Synlett 13:1371
44. Núñez MG, García P, Moro RF, Díez D (2010) Tetrahedron 66:2089
45. Apers S, Vlietinck A, Pieters L (2003) Phytochem Rev 2:201
46. Nising CF, Bräse S (2008) Chem Soc Rev 37:1218
47. Hintermann L (2010) Top Organomet Chem 31:123
48. Nising CF, Bräse S (2012) Chem Soc Rev 41:988
49. Hartmann E, Vyas DJ, Oestreich M (2011) Chem Commun 47:7917
50. Fukata Y, Miyaji R, Okamura T, Asano K, Matsubara S (2013) Synthesis 45:1627
51. Fuwa H, Ichinokawa N, Noto K, Sasaki M (2012) J Org Chem 77:2588
52. Merschaert A, Delbeke P, Daloze D, Dive G (2004) Tetrahedron Lett 45:4697
53. Saito N, Ryoda A, Nakanishi W, Kumamoto T, Ishikawa T (2008) Eur J Org Chem 2008:2759
54. Gioia C, Fini F, Mazzanti A, Bernardi L, Ricci A (2009) J Am Chem Soc 131:9614
55. Tokunou S, Nakanishi W, Kagawa N, Kumamoto T, Ishikawa T (2012) Heterocycles 84:1045
56. Hintermann L, Ackerstaff J, Boeck F (2013) Chem Eur J 19:2311
57. Malerich JP, Hagihara K, Rawal VH (2008) J Am Chem Soc 130:14416
58. Almaşi D, Alonso DA, Gómez-Bengoa E, Nájera C (2009) J Org Chem 74:6163
59. Zhang L, Lee M-M, Lee S-M, Lee J, Cheng M, Jeong B-S, Park H-G, Jew S-S (2009) Adv Synth Catal 351:3063
60. Hammett LP (1937) J Am Chem Soc 59:96
61. Kawato Y, Chaudhary S, Kumagai N, Shibasaki M (2013) Chem Eur J 19:3802
62. Latif Z, Hartley TG, Rice MJ, Waigh RD, Waterman PG (1998) J Nat Prod 61:614
63. Ceccarelli S, De Vellis P, Scuri R, Zanarella S, Brufani M (1993) J Heterocycl Chem 30:679
64. Bitto A, Minutoli L, Squadrito F, Polito F, Altavilla D (2007) Mini Rev Med Chem 7:339
65. Lauret C (2001) Tetrahedron Asymmetry 12:2359
66. Nielsen LPC, Jacobsen EN (2006) In: Yudin AK (ed) Aziridines and epoxides in organic synthesis. Wiley, Weinheim, p 229
67. Adolfsson H (2004) In: Bäckvall J-E (ed) Modern oxidation methods. Wiley, Weinheim, p 21
68. Kinoshita T, Okada S, Park S-R, Matsunaga S, Shibasaki M (2003) Angew Chem Int Ed 42:4680
69. Nishikawa Y, Yamamoto H (2011) J Am Chem Soc 133:8432
70. Russo A, De Fusco C, Lattanzi A (2012) ChemCatChem 4:901
71. Russo A, Galdi G, Croce G, Lattanzi A (2012) Chem Eur J 18:6152
72. Palumbo C, Mazzeo G, Mazziotta A, Gambacorta A, Loreto MA, Migliorini A, Superchi S, Tofani D, Gasperi T (2011) Org Lett 13:6248
73. Chouhan M, Pal A, Sharma R, Nair VA (2013) Tetrahedron Lett 54:7119
74. Uchida R, Imasato R, Shiomi K, Tomoda H, Ōmura S (2005) Org Lett 7:5701
75. Kobayashi Y, Harayama T (2009) Org Lett 11:1603

Top Heterocycl Chem (2017) 50: 95–128
DOI: 10.1007/7081_2015_171

Published online: 14 October 2015

# Chiral Guanidines in Michael Reactions

**Carmen Nájera and Miguel Yus**

**Abstract** Enantioselective Michael additions of carbon and heteronucleophiles to different electrophilic olefins (α,β-unsaturated carbonyl compounds, esters, thioesters, amides, nitriles and nitro compounds) have been efficiently achieved by using chiral acyclic or cyclic guanidines as organocatalysts. They act not only as superbases but also through a hydrogen bond activation mode either of the nucleophile or of the electrophile. By this methodology, a series of polyfunctionalized molecules can be easily prepared in an enantioselective manner.

**Keywords** Carbon and heteronucleophiles · Chiral guanidines · Electrophilic olefins · Michael additions

## Contents



C. Nájera (✉) and M. Yus
Facultad de Ciencias, Departamento de Química Orgánica, Centro de Innovación en Química Avanzada (ORFEO-CINQA), Universidad de Alicante, Apdo. 99, 03080 Alicante, Spain
e-mail: cnajera@ua.es

## Abbreviations

| | |
|---|---|
| Ac | Acetyl group |
| Ar | Aryl group |
| Bn | Benzyl group |
| Boc | *tert*-Butoxycarbonyl group |
| $Bu^n$ | *n*-Butyl group |
| $Bu^t$ | *tert*-Butyl group |
| cat | Catalyst |
| Cy | Cyclohexyl group |
| DCM | Dichloromethane |
| DMF | Dimethylformamide |
| DMSO | Dimethyl sulfoxide |
| dr | Diastereomeric ratio |
| ee | Enantiomeric excess |
| Et | Ethyl group |
| LUMO | Lowest unoccupied molecular orbital |
| Me | Methyl group |
| NBS | *N*-Bromosuccinimide |
| NMR | Nuclear magnetic resonance |
| Ph | Phenyl group |
| PhMe | Toluene |
| $Pr^i$ | Isopropyl group |
| $Pr^n$ | *n*-Propyl group |
| rt | Room temperature |
| TBDMS | *tert*-Butyldimethylsilyl group |
| TBDPS | *tert*-Butyldiphenylsilyl group |
| TBS | *tert*-Butyldimethylsilyl group |
| TFA | Trifluoroacetic acid |
| THF | Tetrahydrofuran |
| TMS | Trimethylsilyl group |
| Ts | 4-Methylphenylsulfonyl group |

## 1 Introduction

Guanidines play an important role in organic synthesis [1] especially due to their ability to act as superbases [2] owing to the resonance stabilization of their conjugated acids. Taking advantage of this behaviour, it is expected that guanidines can catalyse various types of base-mediated organic reactions. This possibility can be classified in the frame of the so-called organocatalysis that includes transformations catalysed by a small organic molecule, which in the last few years has attracted enormous interest in synthetic organic chemistry because they are experimentally simple and show low toxicity and high efficiency and selectivity [3–6].

One especially interesting case from a synthetic point of view is the use of chiral guanidines as catalysts in enantioselective transformations [7–10].

In this chapter, we will consider the guanidine-catalysed enantioselective addition of different carbon, oxygen, nitrogen, sulphur or phosphorous nucleophiles to electrophilic olefins, such as α,β-unsaturated carbonyl compounds, esters and lactones, amides or nitro derivatives [11]. In the last part, some cyclization processes will be also studied. In all cases, an especial attention will be paid to the synthetic applications of these reactions to the enantioselective preparation of polyfunctionalized organic compounds.

## 2 Carbon Nucleophiles

The most studied nucleophiles to be added to electrophilic olefins are the carbon-derived ones, which together with the formation of a new carbon–carbon bond generates at least one new stereogenic centre, depending on the structure of both the nucleophile and the Michael acceptor. In this section, the guanidine-catalysed addition of different nucleophiles to electrophilic olefins is classified according to the last ones.

### *2.1 Addition to α,β-Unsaturated Carbonyl Compounds*

The addition of dibenzyl malonates (**2**) to cyclopentenone (**1**) was achieved in an enantioselective manner in 2005 by using the guanidine **I** as a catalyst, so the expected products **3** were obtained with moderate asymmetric induction [12] (Scheme 1).

The same catalyst **I** and the $C_2$-symmetrical one **II** were previously used to prepare the structural core of (−)-huperzine A as it is illustrated in Scheme 2. Starting from compounds **4** and **5**, and using the catalyst **II** in stoichiometric amount, a tandem Michael addition followed by an intramolecular aldol reaction allowed the formation of compound **6**. Final transformation into the related

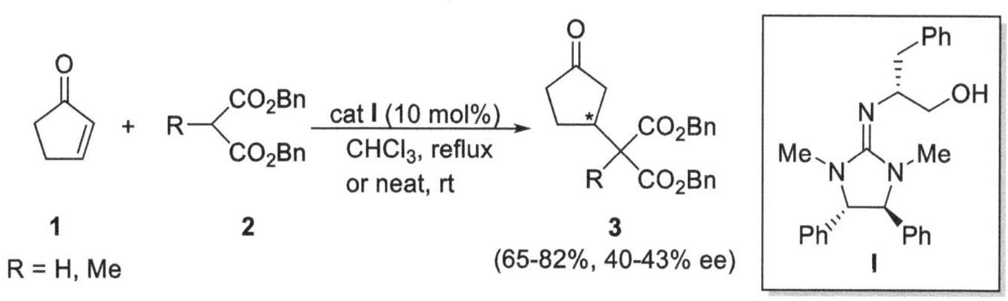


**Scheme 1** Conjugate addition of malonates to cyclopentenone

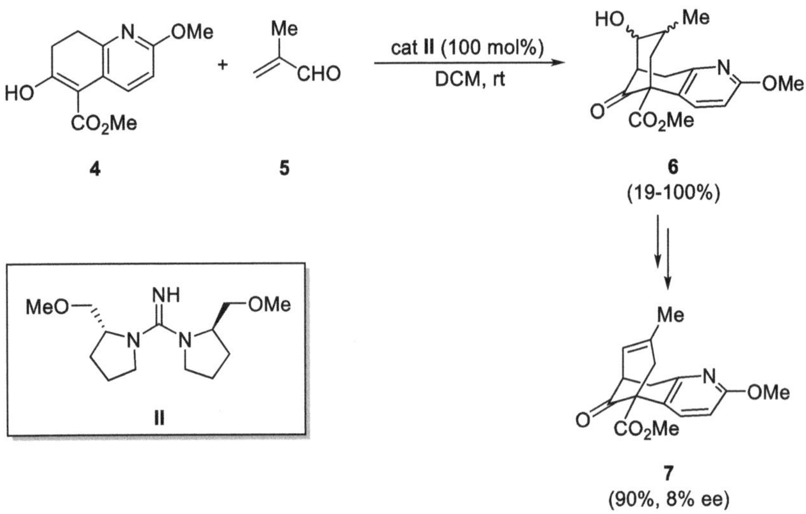


**Scheme 2** Synthesis of the (−)-huperzine A core

intermediate **7**, rather close to the mentioned natural product, worked with poor optical yield [13, 14].

More rigid bicyclic guanidines of type **III** have been widely used in asymmetric organocatalysis. In 2006, the addition of nitro compounds **8** to chalcone (**10**), or dimethyl malonate (**9**) to the diketone **11** catalysed by **III**, was reported to give the expected products **12** or **13** with poor to moderate yields and enantioselectivities [15] (Scheme 3).

The guanidine **IIId** was successfully used in the enantioselective addition of dithiomalonates and β-keto thioesters **15** to cyclopentenone (**1**) and cyclohexanone (**14**) or α,β-unsaturated β-keto esters **16** to give the expected products **17** and **18**, respectively, with both good yields and stereoselectivities [16] (Scheme 4). The organocatalyst **IIId** is one of the most studied guanidines in enantioselective Michael additions.

A related chemistry was developed using the same catalyst **IIId** for the conjugate addition of the dithiomalonate **15b** to a series of enones bearing an amide functionality **19**, giving the expected products **20** with variable yields and good enantioselectivities [17] (Scheme 5). The same approach was studied using differently substituted 1,4-enediones as Michael acceptors in order to investigate the regioselectivity of the process.

In the case of starting from α-fluoro β-keto esters **21** and some amido enones **19**, the same reaction gave the corresponding fluorinated products **22** under similar reaction conditions and with excellent yields as well as enantio- and diastereoselectivities [18] (Scheme 6). The observed diastereo- and enantioselectivities were explained by considering the optimized (B3LYP/6-31G*) geometries of the four possible transition states leading to the obtained products.

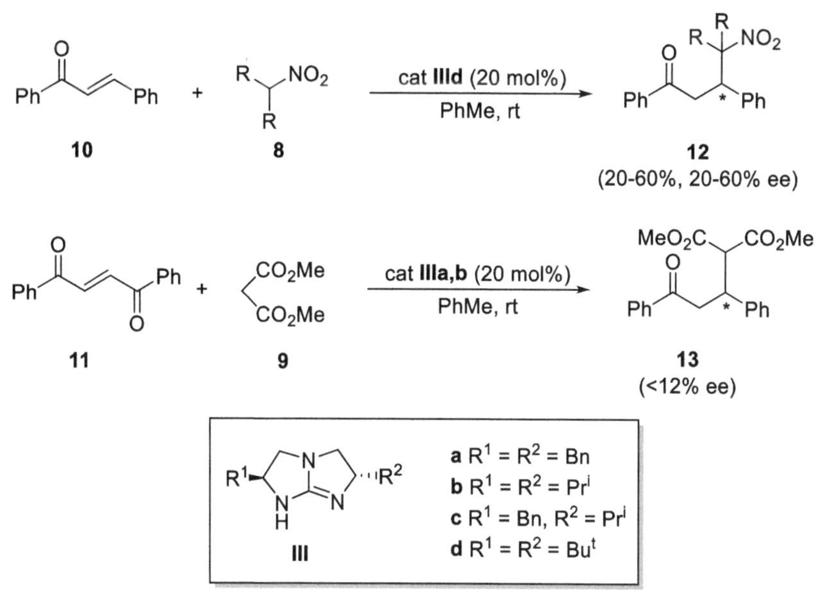


**Scheme 3** Addition of nitroalkanes or dimethyl malonate to α,β-unsaturated ketones

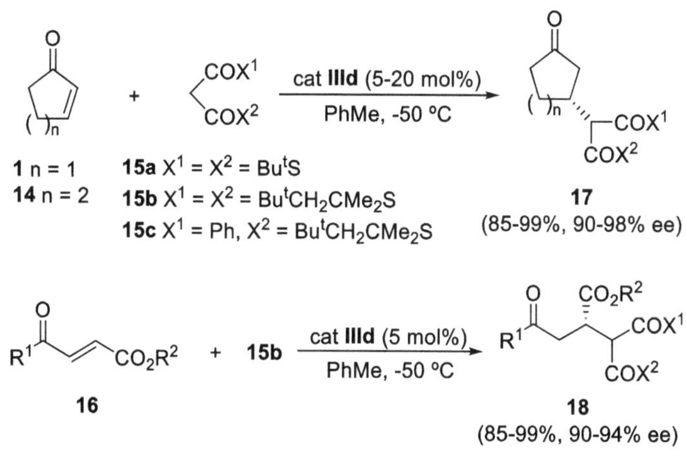


$R^1$ = Ph, 3-$MeOC_6H_4$, 2-naphthyl, 4-$Bu^tC_6H_4$, 4-$PhC_6H_4$, 3,5-$Me_2C_6H_3$, 2-furyl, 4-$ClC_6H_4$, 4-$NCC_6H_4$

$R^2$ = Me, Et, Bn

$X^1 = X^2 = Bu^tCH_2CMe_2S$

**Scheme 4** Addition of dithiomalonates or β-keto thioesters to enones

An interesting rate acceleration effect of triethylamine (the reaction rate can be increased up to 1,000 times) was observed for the Michael addition of 1,3-diketones, β-keto esters or malonates **23** to cyclopentenone (**1**) in the presence

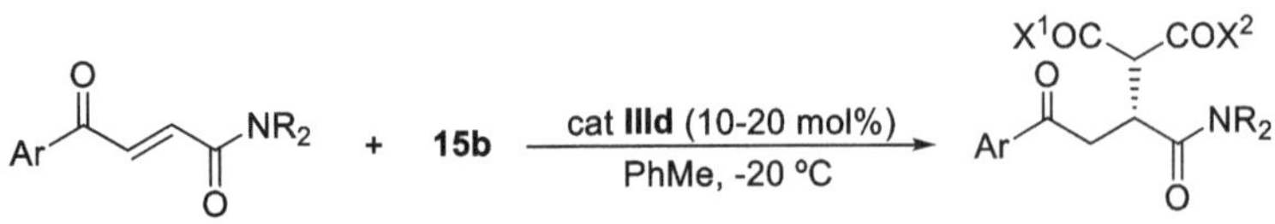


19

20
(24-99%, 59-97% ee)

Ar = Ph, 4-NCC$_6$H$_4$, 4-FC$_6$H$_4$, 4-ClC$_6$H$_4$, 4-BrC$_6$H$_4$, 4-PhC$_6$H$_4$, 4-furyl, 2-naphthy, 3-NCC$_6$H$_4$, 3-MeOC$_6$H$_4$, 2-O$_2$NC$_6$H$_4$
NR$_2$ = NEt$_2$, N(CH$_2$)$_4$, N(CH$_2$)$_5$, N(CH$_2$CH$_2$)$_2$O, N(CH$_2$)$_3$CO, N(CH$_2$)$_2$OCO
X$^1$ = X$^2$ =Bu$^t$CH$_2$CMe$_2$S

**Scheme 5** Addition of the dithiomalonate **15b** to enones **19**

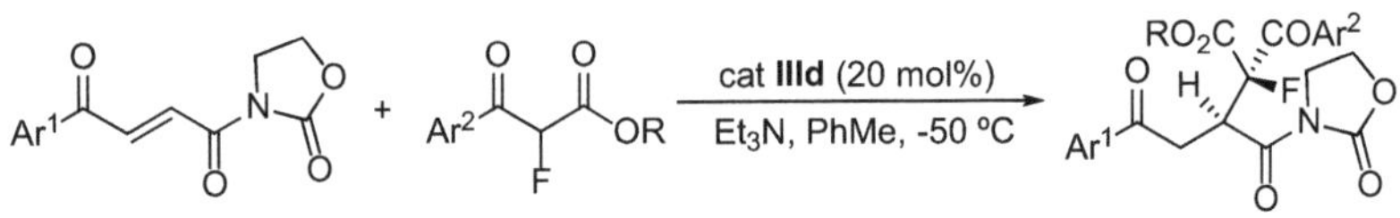


19 21 22
(91-94%, 99:1 dr, 92-96% ee)

Ar$^1$ = 4-ClC$_6$H$_4$, 4-BrC$_6$H$_4$
Ar$^2$ = 3-MeOC$_6$H$_4$, 3,5-Me$_2$C$_6$H$_3$
R = Me, Et

**Scheme 6** Addition of fluoro keto esters **21** to the enone **19**

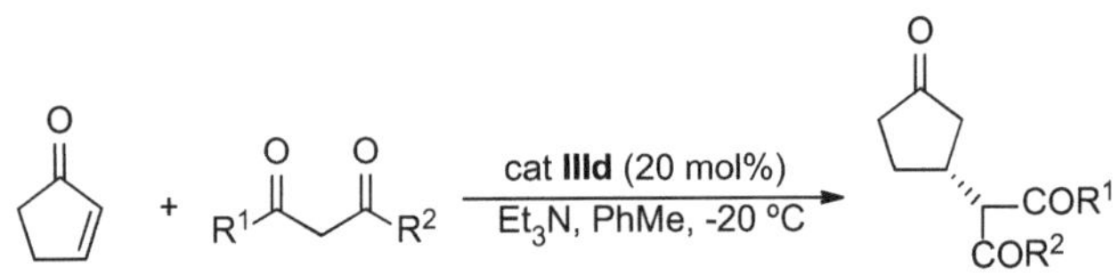


1 23 24
(84-99%, 78-94% ee)

R$^1$ = Ph, 3-MeC$_6$H$_4$, 4-F$_3$CC$_6$H$_4$, 4-ClC$_6$H$_4$, 3-ClC$_6$H$_4$, 4-O$_2$NC$_6$H$_4$, OMe, OEt, OBn, OPr$^i$
R$^2$ = Ph, OMe, OEt, OBn, OPr$^i$

**Scheme 7** Addition of 1,3-dicarbonyl compounds to cyclopentenone

of catalytic amounts of the guanidine **IIId**, so the corresponding products **24** were isolated with excellent results [19] (Scheme 7).

As it was shown in Scheme 3, nitro compounds are adequate nucleophiles for the conjugate addition to electrophilic olefins. In this case, the addition of nitroalkanes **8** to methyl vinyl ketone (**25**) in the presence of catalytic amounts of the chiral bicyclic guanidine **IV** gave the expected Michael adducts **26** with poor enantio-selectivities [20] (Scheme 8).

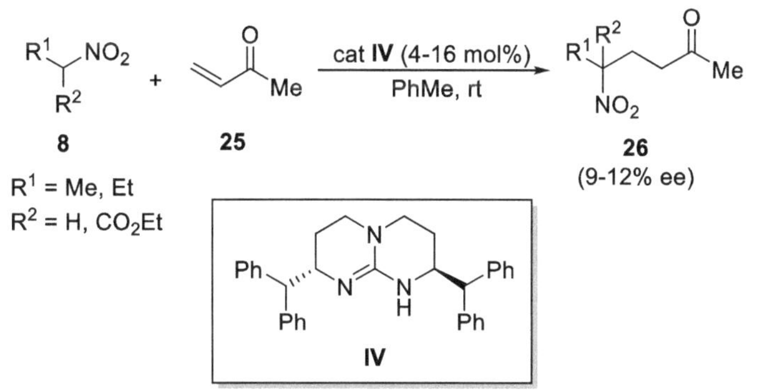


**Scheme 8** Addition of nitroalkanes to methyl vinyl ketone

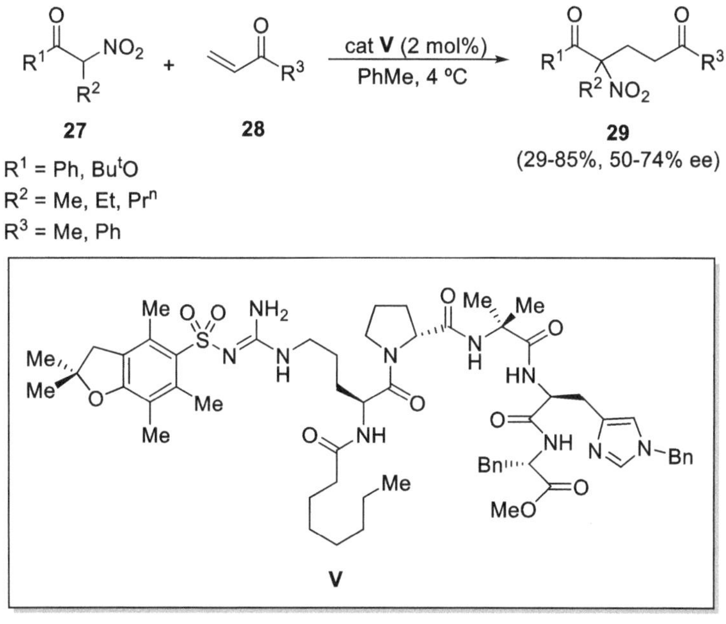


**Scheme 9** Addition of nitro compounds **27** to vinyl ketones **28**

In the case of starting from α-nitro carbonyl compounds **27**, the reaction with different enones **28** yielded the corresponding products **29** under the catalytic action of the peptide **V** in which a *N*-sulfonyl guanidine moiety derived from arginine is present [21] (Scheme 9). The reaction of α-nitro ketones having an aliphatic substituent ($R^1$=alkyl) or an α-nitro ethyl ester ($R^1$=OEt) with methyl vinyl ketone gave racemic products. The catalyst was prepared through Fmoc-solid-phase peptide synthesis and was directly used after cleavage from the resin.

The two not so complicated (as compared to **V**) peptides **VI** and **VII** also derived from arginine have been explored as catalysts in the addition of 2-nitropropane (**8**)

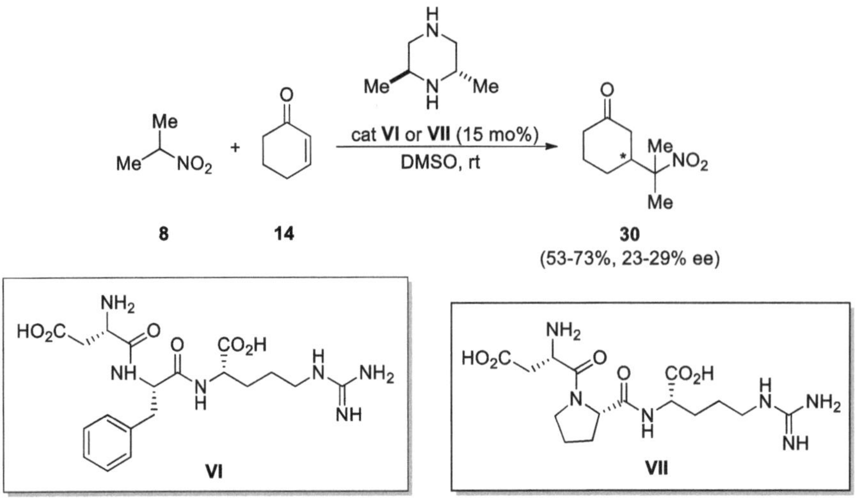

Scheme 10 Addition of 2-nitropropane to cyclohexenone

to cyclohexenone (**14**) in the presence of *trans*-2,5-dimethylpiperazine giving modest results of the corresponding product **30** [22] (Scheme 10).

A tandem Henry–Michael reaction involving 6-oxo enals **31** and nitromethane (**8**) has been achieved in an asymmetric version using the guanidine **VIII**. The chiral substituted cyclohexanols **32** were obtained with excellent results in terms of yields and dr and ee values [23] (Scheme 11). According to the already reported mechanism and control experiments, the enantioselectivity of the reaction is generated in the intramolecular Michael addition step, giving finally the thermodynamically most stable product.

An interesting nucleophile for the conjugate addition is the glycine α-imino ester derivative **33** which was allowed to react with methyl vinyl ketone (**25**) in the presence of a catalytic amount of the chiral guanidine **IX**, giving the expected product **34** with excellent chemical yield but poor enantioselectivity [24] (Scheme 12).

From different bicyclic guanidines having a five- and six-membered ring junction, compound **VIII** turned out to be an excellent catalyst for the Michael addition of oxazolones **35** to the ynone **36**, giving the corresponding *Z*-derivatives **37** as the major products in high yields and stereoselectivities [25] (Scheme 13). It is likely that one face of the enolate anion in the intermediate **38** is shielded by the 5*H*-oxazole ring because the electron-enriched π orbital of the enolate interacts with the electron-deficient carbon atom at the 2-position of the oxazole ring. Therefore, intermediate **38** most likely undergoes protonation from the other face giving the thermodynamically unstable *Z*-isomer with *R*-configuration.

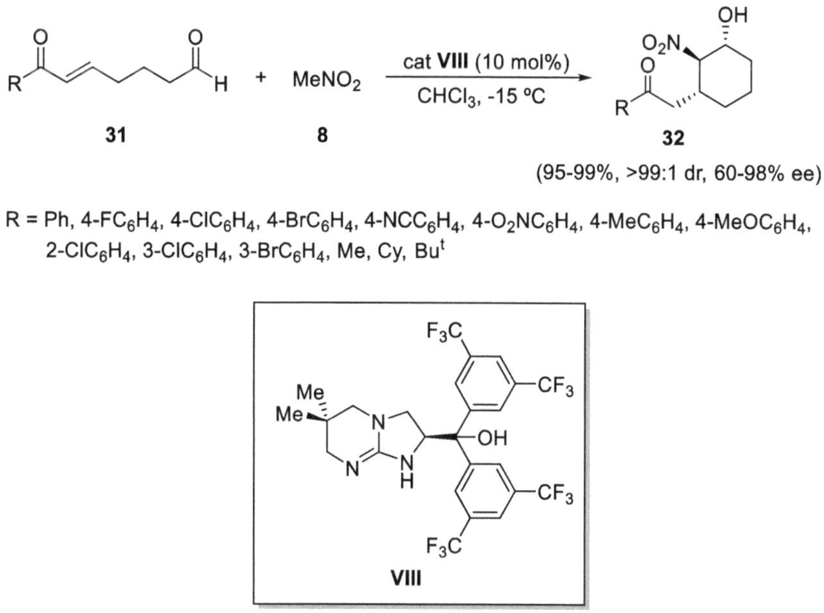


**Scheme 11** Cyclization of 6-oxo enals **31** with nitromethane

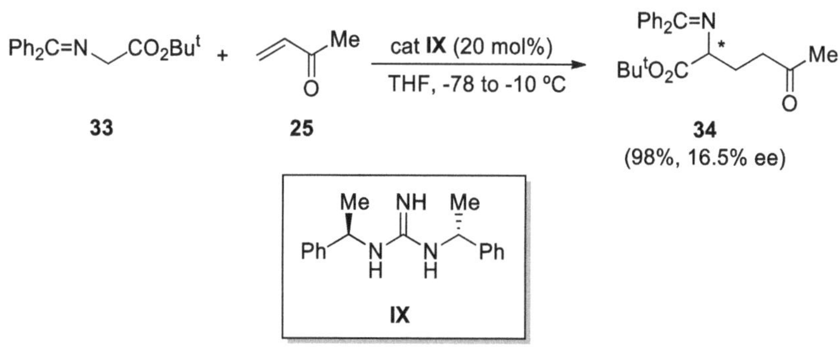


**Scheme 12** Addition of the glycine derivative **33** to methyl vinyl ketone

## 2.2 *Addition to α,β-Unsaturated Esters and Thioesters*

The addition of dimethyl malonate (**9**) to dimethyl fumarate (**39**) in the presence of guanidines **IIIa** or **IIIb** as catalysts only gave a low level of enantioselectivity in the products **40** [15] (Scheme 14).

The α,β-unsaturated γ-butyrolactam **41** added to alkylidene malonates **42** with the guanidine **X** acting as the appropriate catalyst, so the expected products **43** were obtained with good yields and diastereo- and enantioselectivity [26] (Scheme 15). In this case, a dual activation of the generated dienolate (the basic guanidine

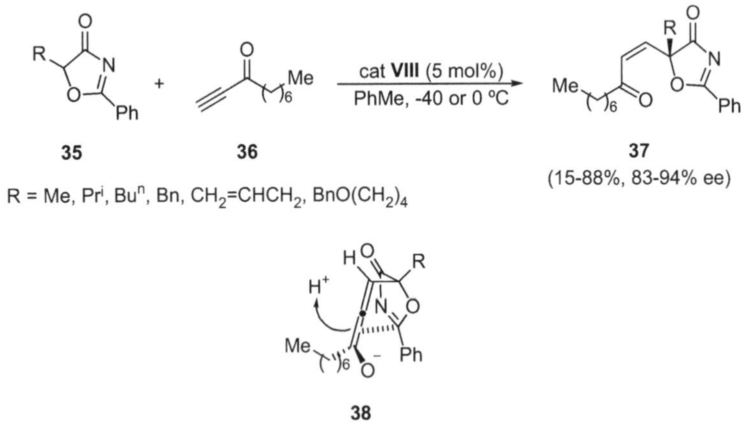


**Scheme 13** Addition of oxazolones **35** to the ynone **36**

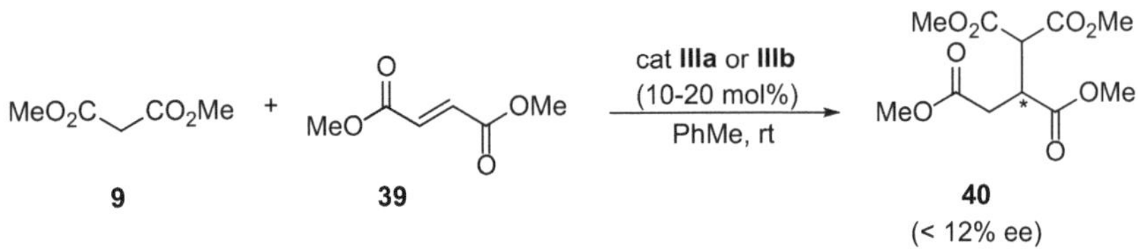


**Scheme 14** Addition of dimethyl malonate to dimethyl fumarate

accelerates its formation) and the diester takes place through a network of hydrogen bonds of NH groups. Therefore, the desired product could be obtained by the *Re*-face attack of the activated alkylidene malonate, as it is shown in the topicity **44**.

Benzophenoneimine glycinates **33** could be added to different acrylates **45** in the presence of guanidines **IX** and **XI–XIII** to yield the corresponding Michael adducts **46** with good yields but low enantioselectivities [24] (Scheme 16). A possible mechanism would involve an interaction of the deprotonated imino ester **33** with the chiral guanidine (for instance, **IX**) giving a complex **47** which reacts with the acrylate **45** to deliver the product **46**.

The same reaction as shown in Scheme 16 was also carried out by means of other types of chiral guanidines, namely, catalysts **I** or **XIV**, giving the expected products **46** with variable yields and good to excellent enantioselectivities [27] (Scheme 17). The guanidine **XIVa** gave the best results under neat conditions.

Dithiomalonate **15** was able to be added to the butenolide **48** in a reaction promoted by the guanidine **IIId**, affording the corresponding product **49** with both high yields and enantioselectivities [16] (Scheme 18).

The enantioselective addition of oxazolones **35** to the acetylenic thioester **50** catalysed by the guanidine **IX** has also been reported. The expected *Z*-adducts **51** with *R*-configuration were obtained with variable results [25] (Scheme 19). The explanation of the obtained enantioselection has been already shown in Scheme 13.

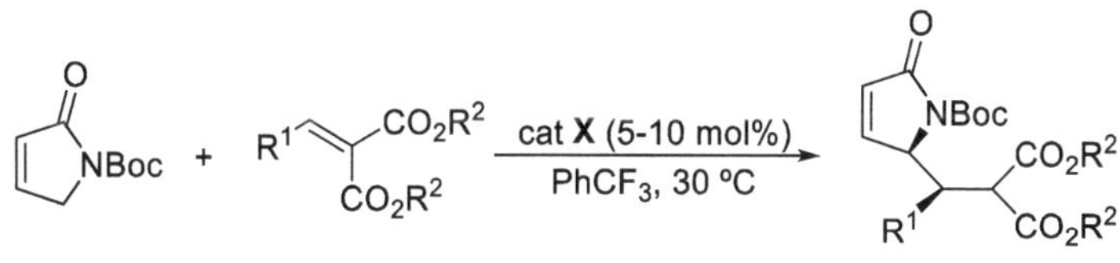


**41** **42** **43**
(64-89%, 78-94% ee)

$R^1$ = Ph, 4-$FC_6H_4$, 2-$ClC_6H_4$, 3-$ClC_6H_4$, 4-$ClC_6H_4$, 3-$BrC_6H_4$, 4-$BrC_6H_4$, 3-$MeC_6H_4$, 4-$MeC_6H_4$, 3-$MeOC_6H_4$, 4-$MeOC_6H_4$, 4-$PhC_6H_4$, 3-$PhOC_6H_4$, 2-thienyl, 3-thienyl, 2-naphthyl, (*E*)-PhCH=CH, Cy
$R^2$ = Me, Et, Bn

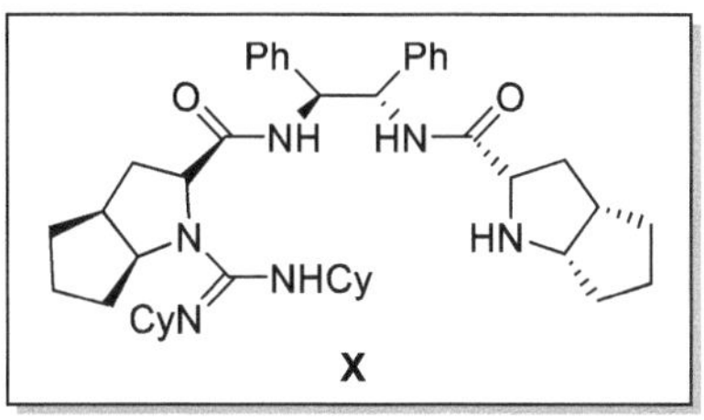


**X**

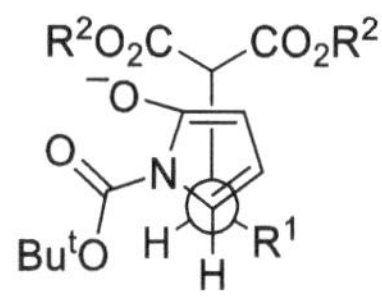


**44**

**Scheme 15** Addition of the lactam **41** to alkylidene malonates

## 2.3 *Addition to α,β-Unsaturated Amides*

Almost all reactions involving enantioselective nucleophilic additions to α,-β-unsaturated amides catalysed by chiral guanidines take place on *N*-substituted maleimides. Thus, several 1,3-dicarbonyl compounds **15**, such as 1,3-diketones, β-keto esters, β-keto thioesters or dithiomalonates were added to *N*-alkyl maleimides **52** under catalysis with the guanidine **IIId**, affording the corresponding succinimides **53** with good yields and enantioselectivities [16] (Scheme 20).

The same type of maleimides **52** was adequate Michael acceptors for 1,3-dicarbonyl compounds **15**, namely, 1,3-diketone, β-keto esters or dimethyl malonates using the guanidine **XV** derived from 2-aminobenzimidazole, so enantioenriched products **53** were obtained [28, 29] (Scheme 21). For 1,3-diketones or β-keto esters, the best results were obtained using trifluoroacetic acid as cocatalyst. Computational and NMR studies support a hydrogen-bonding activation role of the catalyst of both the nucleophile and the electrophile as the origin of the stereoselectivity of the process.

As it was commented in Scheme 7, an acceleration of the reaction rate was also observed by using triethylamine in the conjugate addition of β-keto esters and dimethyl malonates **23** to *N*-substituted maleimides **52**. Products **53** were obtained with excellent yields and enantioselectivities [19] (Scheme 22). Cyclic β-keto esters derived from

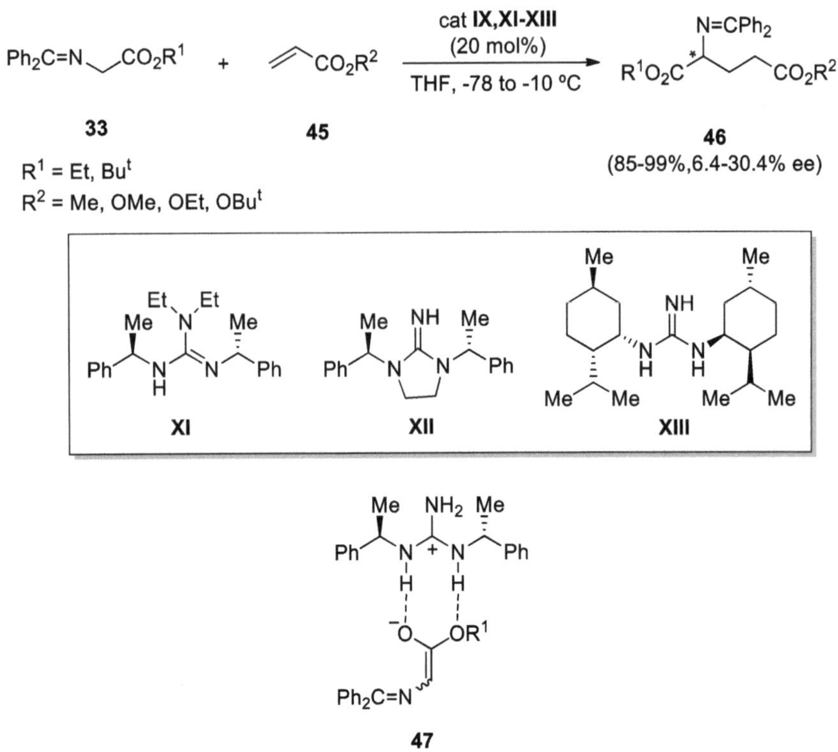


**Scheme 16** Addition of the glycine derivative **33** to acrylates

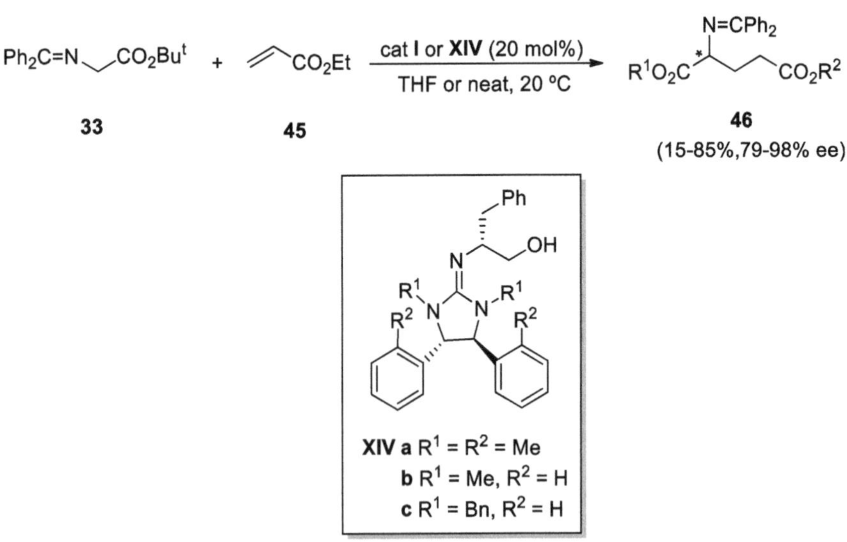


**Scheme 17** Addition of the glycine derivative **33** to ethyl acrylate

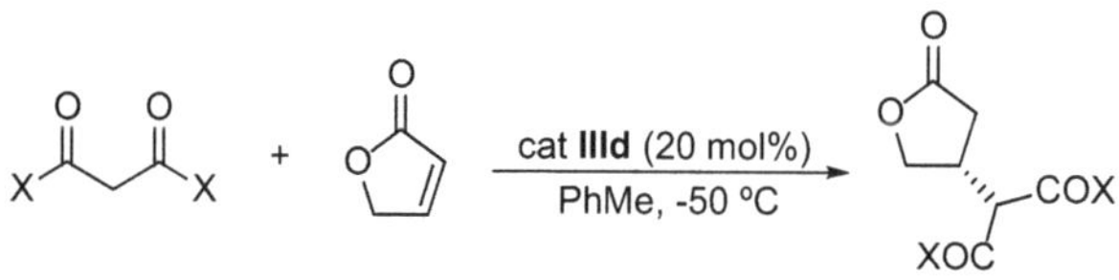


15 48 49

X = Bu$^t$CH$_2$CMe$_2$S (85%, 96% ee)

**Scheme 18** Addition of dithioesters to the butenolide **48**

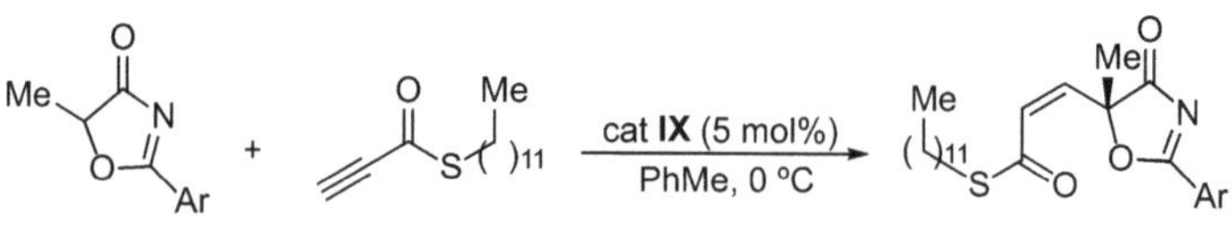


35 50 51

Ar = Ph, 4-F$_3$CC$_6$H$_4$, 4-Pr$^i$C$_6$H$_4$ (33-88%, 38-91% ee)

**Scheme 19** Addition of oxazolones **35** to the acetylenic thioester **50**

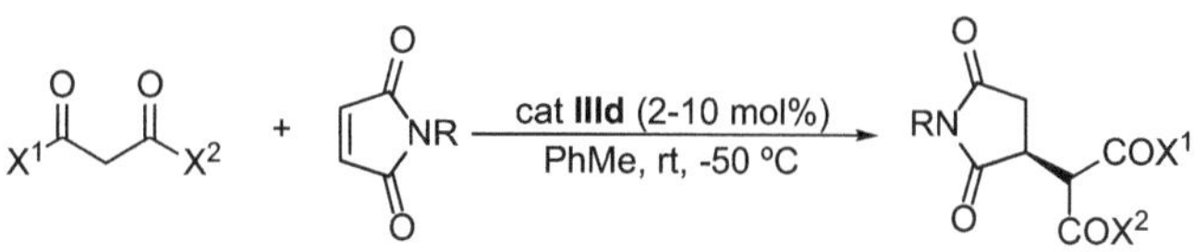


15 52 53

(87-99%, 78-97% ee)

X$^1$ = Me, Ph, 4-MeOC$_6$H$_4$, Bu$^t$S
X$^2$ = EtO, Bu$^t$S, Bu$^t$CH$_2$CMe$_2$S
R = Me, Et, *n*-C$_6$H$_{13}$, CyCH$_2$, Bn

**Scheme 20** Addition of 1,3-dicarbonyl compounds **15** to maleimides

cyclopentanone (namely, 2-methoxy- and ethoxycarbonylcyclopentanone) gave also very good results in the same reaction with *N*-ethyl maleimide (**52**, R$^3$=Et).

The guanidine **IIId** has shown to be very effective in catalysing the addition of fluorinated derivatives of type **21** to *N*-substituted maleimides **52** in order to prepare the expected products **53** with excellent results [18] (Scheme 23). Computational studies could explain properly the obtained results concerning diastereo- and enantioselectivity.

The addition of substituted anthrones **54** to *N*-substituted maleimides **52** was achieved in an enantioselective manner using guanidines **IIIa** and **VIII**, so compounds **55** were prepared with good to excellent results [30, 31] (Scheme 24). The formation of the Michael adduct, instead of the cycloadduct (see Sect. 7), was attributed either to the retroaldol ring opening of the cycloadduct under basic conditions or directly through a Michael reaction.

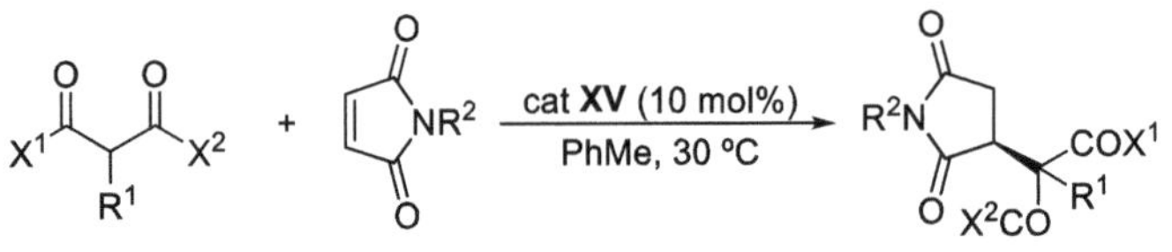

15 52 53

(87-99%, 85:15-95:5 dr,78-97% ee)

$X^1$ = Me, Ph, MeO
$X^2$ = Me, MeO
$R^1$ = H, Me
$R^2$ = H, Me, Et, Bn, Ph, 4-$BrC_6H_4$, 4-$AcC_6H_4$

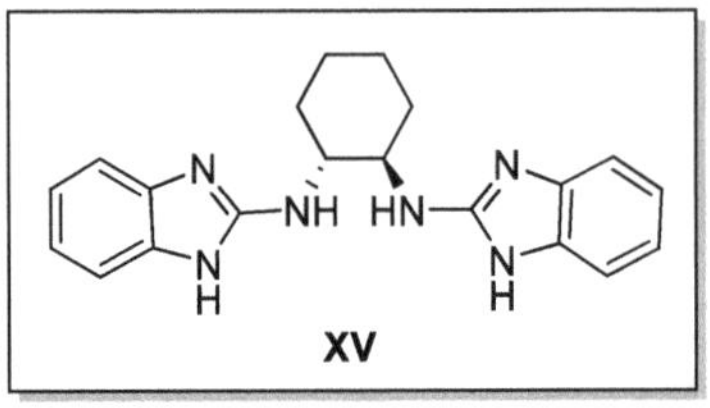

**Scheme 21** Addition of 1,3-dicarbonyl compounds to maleimides

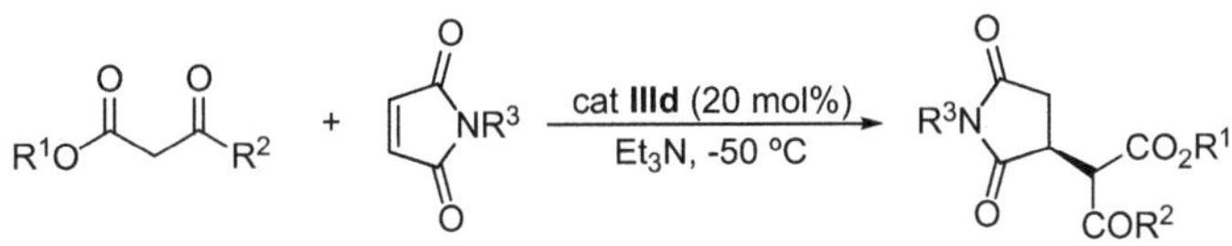

23 52 53

(88-99%, 1:1 dr, 86-93% ee)

$R^1$ = Me, Et
$R^2$ = Ph, 3-$MeC_6H_4$, 4-$ClC_6H_4$, 4-$MeOC_6H_4$, 2-furyl
$R^3$ = Me, Et, Cy

**Scheme 22** Addition of 1,3-dicarbonyl compounds to maleimides

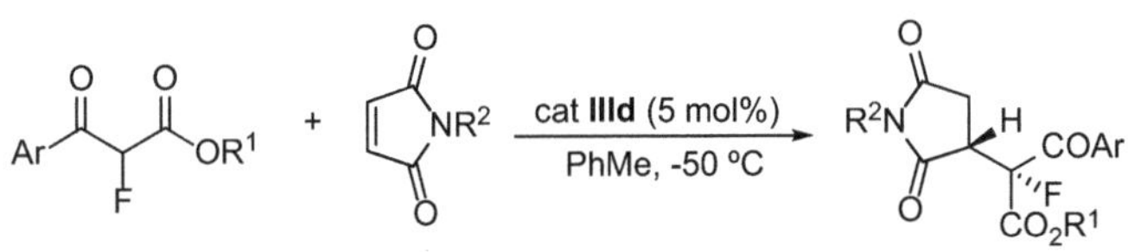

21 52 53

(80-99%, 98:2->99:1 dr, 83-97% ee)

$R^1$ = Me, Et
$R^2$ = Me, Et, Cy, Bn, n-$C_6H_{13}$, $Bu^t$
Ar = Ph, 4-$O_2NC_6H_4$, 4-$F_3CC_6H_4$, 4-$ClC_6H_4$, 4-$BrC_6H_4$, 4-(4-$BrC_6H_4$)$C_6H_4$, 3-$MeOC_6H_4$, 4-$MeOC_6H_4$, 4-$BnOC_6H_4$, 3,5-$Me_2C_6H_3$, 2-naphthyl, 2-thienyl

**Scheme 23** Addition of fluorinated compounds **21** to maleimides

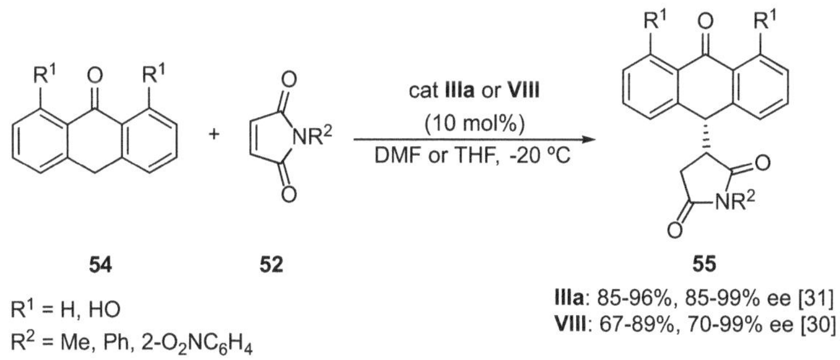


**Scheme 24** Addition of anthrones **54** to maleimides

Hindered α-disubstituted aldehydes **56** were efficiently added to *N*-substituted maleimides **52** in a reaction catalysed by the guanidine **XVI**, so the corresponding products **57** were obtained with excellent yields and enantioselectivities [32, 33] (Scheme 25). Using this primary amine–guanidine organocatalyst, the intermediate enamine can be formed, which is able to add even to unsubstituted maleimide. The reaction was carried out in the presence of imidazole as additive and using a mixture of DMF and water as solvents. An opposite enantioselection, compared to that observed using the corresponding thiourea, was obtained, so suggesting a different activation mode in both cases. Theoretical calculations (DFT and M06-2X) suggested that a different hydrogen-bonding coordination pattern between the maleimide (C=O) and the catalyst (NH) is responsible for the observed enantioselectivity.

Special nucleophiles able to be added to *N*-substituted maleimides **52** are the oxindoles **58**, which in the presence of the guanidine **IIId** gave the expected products **59** with excellent yields and enantioselectivities, yet variable diastereoselectivities [34] (Scheme 26). An interesting feature of the process is the generation of a quaternary stereocentre at the 3-position. The obtained results were explained through a transition state based on steric effects.

The only example of a conjugate addition of oxazolones **35** to an acetylenic amide **60** was carried out using the chiral guanidine **IX**. The expected products **61** were prepared with variable yields and high enantio- and diastereoselectivities for the *Z*-products [25] (Scheme 27).

## 2.4 Addition to α,β-Unsaturated Nitriles

Whereas the imino glycinate **33** did not react with acrylonitrile (**62**) in the presence of the guanidine **I** in tetrahydrofuran, when the reaction was performed under neat conditions, the corresponding product **63** was obtained with good yield and moderate enantioselectivity [35] (Scheme 28).

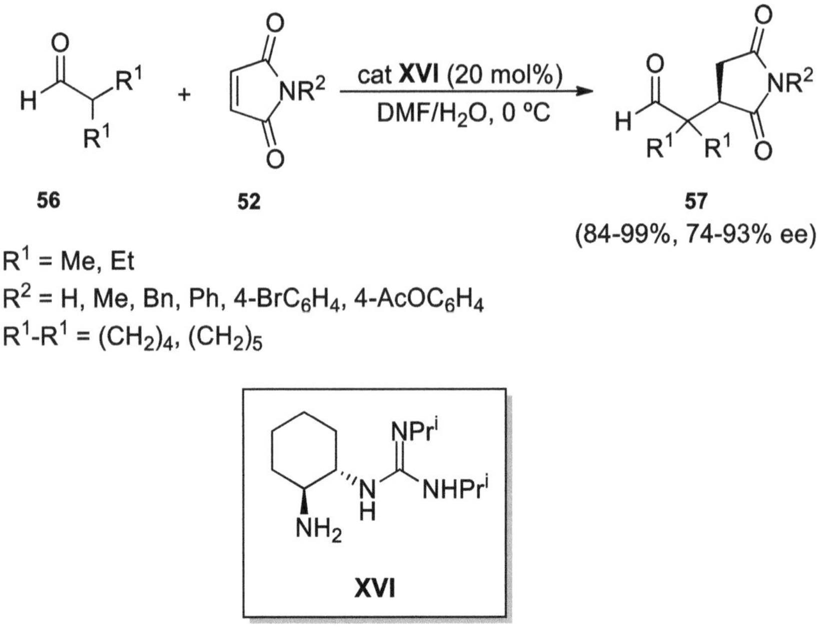


**Scheme 25** Addition of aldehydes to maleimides

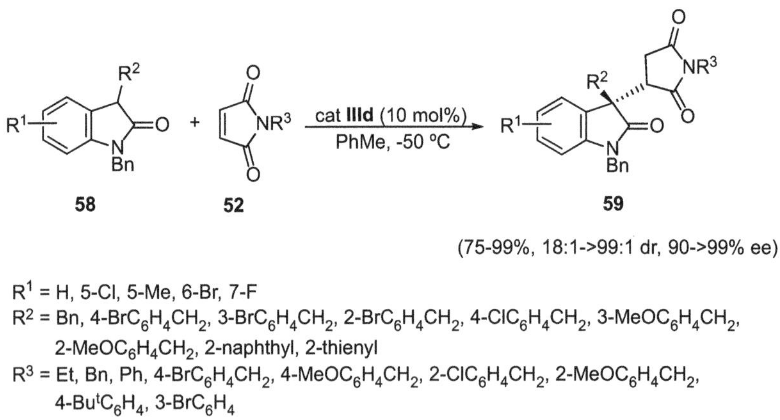


**Scheme 26** Addition of oxindoles **58** to maleimides

## 2.5 *Addition to α,β-Unsaturated Nitro Compounds*

The binaphthyl-derived guanidine **XVII** has been used as an efficient catalyst for the addition of dialkyl malonates **9** to nitroalkenes **64**, so the expected addition products **65** were obtained, in general, with excellent yields and enantioselectivities [36, 37] (Scheme 29). The same process has also been applied to acetyl acetone or

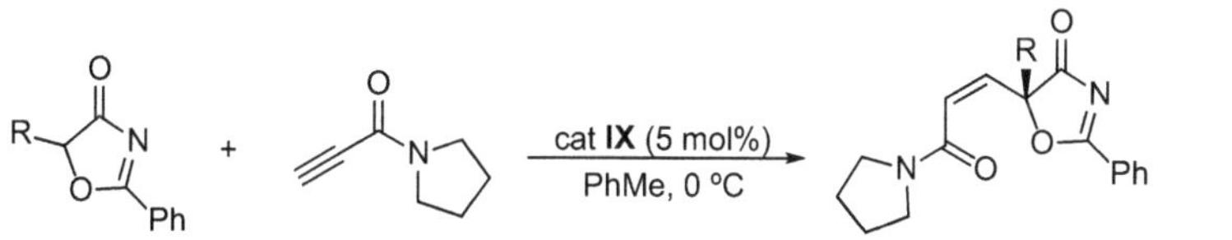


**35** **60** **61**

(43-55%, >99:1 *Z/E*, 94-99% ee)

R = Me, $Pr^i$, $CH_2$=$CHCH_2$, Bn, $BnO(CH_2)_4$

**Scheme 27** Addition of oxazolones **35** to the propiolamide **60**

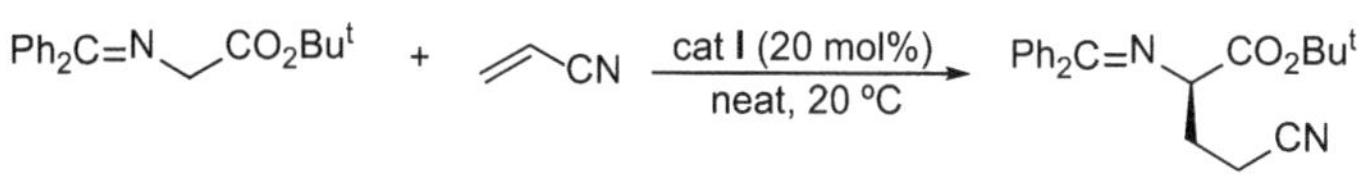


**33** **62** **63**

(79%, 55% ee)

**Scheme 28** Addition of the glycine derivative **33** to acrylonitrile

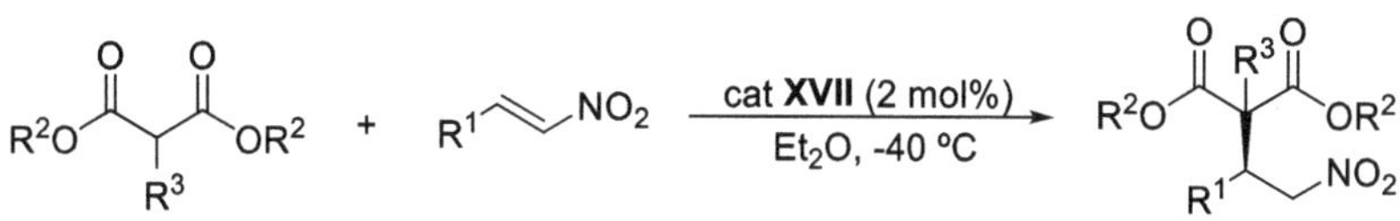


**9** **64** **65**

(86->99%, 87-98% ee)

$R^1$ = $Pr^i$, $Pr^iCH_2$, Cy, 2-$MeOC_6H_4$, 2-$BrC_6H_4$, 2-$O_2NC_6H_4$, 3-$MeOC_6H_4$, 3-$BrC_6H_4$, 4-$MeOC_6H_4$, 4-$ClC_6H_4$, 4-$BrC_6H_4$, 2-furyl, 2-naphthyl

$R^2$ = Me, Et, $Pr^i$

$R^3$ = H, Me

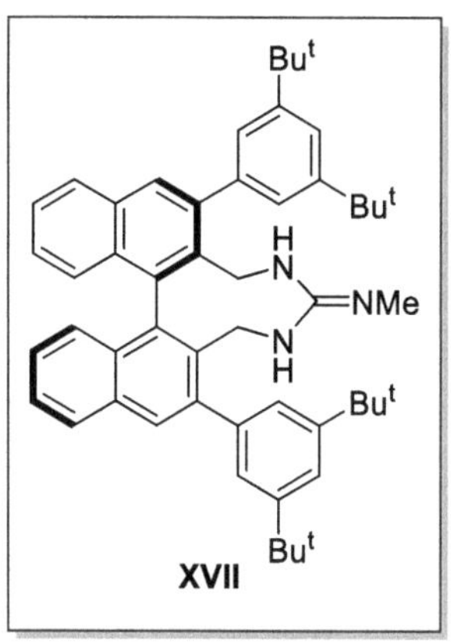


**Scheme 29** Addition of malonates to nitroalkenes

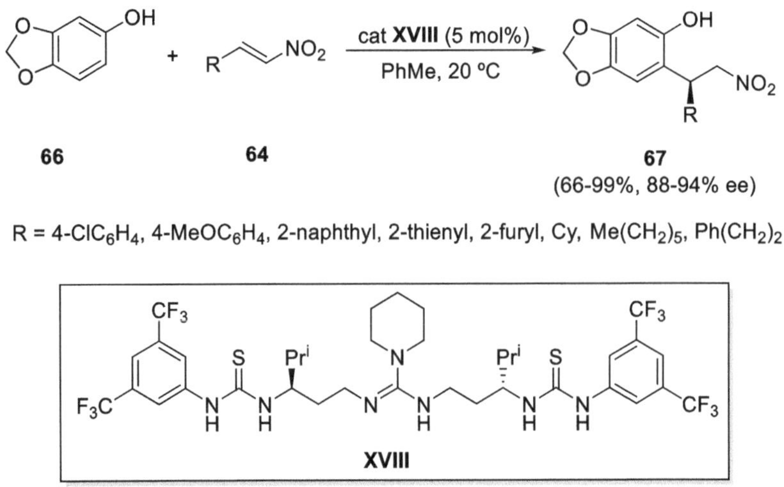


**Scheme 30** Addition of sesamol to nitroalkenes

methyl acetoacetate with similar results. Other guanidines with different aryl groups at the 3 and 3′ positions [Ph, 4-$F_3CC_6H_4$, 3,5-$(F_3C)_2C_6H_3$, 4-$Bu^tC_6H_4$, 3,5-$Ph_2C_6H_3$, 3,5($Bu^t_2C_6H_3$)$C_6H_3$] or at the imine nitrogen ($Pr^n$) gave worse results.

Substituted phenols, for instance, sesamol (**66**) or 1- or 2-naphthol, were added to nitroalkenes **64** in a reaction catalysed by the guanidine **XVIII** (bearing also two thiourea moieties) to yield 2-substituted derivatives **67** in good yields and enantioselectivities [38, 39] (Scheme 30). A working model in which one of the thioureas activates the nitro group and the guanidine moiety activates the nucleophile, both through hydrogen bonding, was proposed to explain the bifunctional character of the catalyst.

Bicyclic guanidines of types **XIX** and **XX** derived from 1,2-diaminocyclohexane have found interesting applications in the isomerization of β-keto acetylenes and can be also used for the enantioselective addition of diethyl malonate to nitrostyrene [40] (see Scheme 29). The reaction used 10 mol% of the catalyst and was conducted in toluene at room temperature to yield the expected products in good yields (62–95%) and enantioselectivities (76–85% ee).

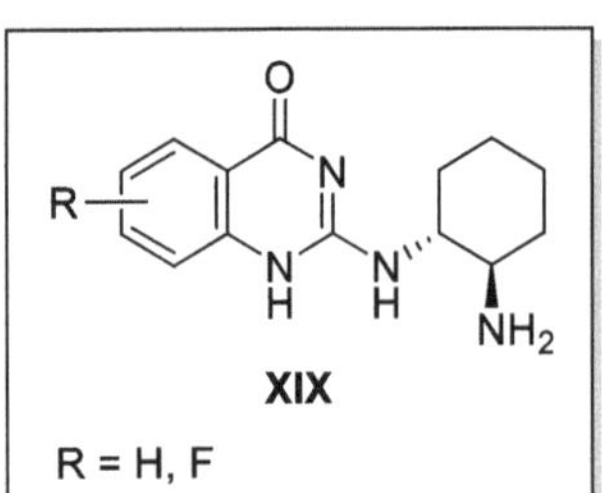


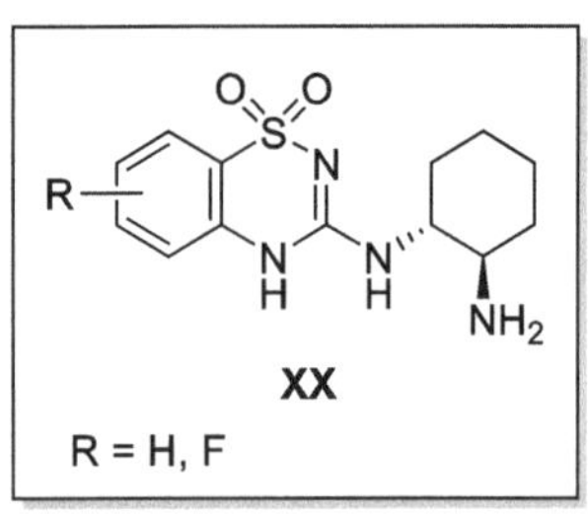

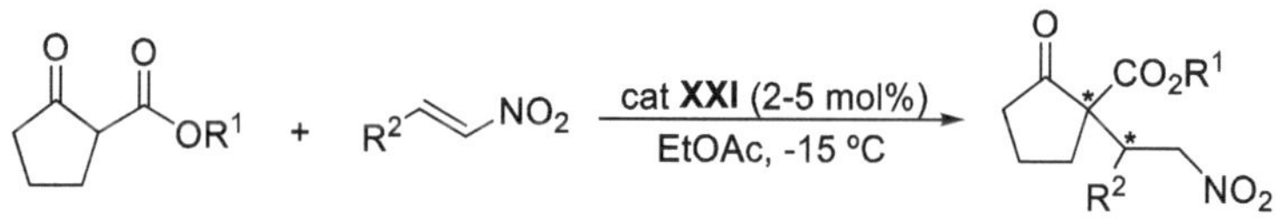


68 64 69

(58-99%, 78:22->99:1 dr, 73-96% ee)

$R^1$ = Et, $Bu^t$, Cy, 1-adamantyl

$R^2$ = Ph, 2-$MeC_6H_4$, 3-$MeC_6H_4$, 4-$MeC_6H_4$, 3-$MeOC_6H_4$, 4-$MeOC_6H_4$, 3,4-$(MeO)_2C_6H_3$, 3,4-$(OCH_2O)C_6H_3$, 2-$ClC_6H_4$, 3-$ClC_6H_4$, 4-$ClC_6H_4$, 2,4-$Cl_2C_6H_3$, 2,6-$Cl_2C_6H_3$, 4-$FC_6H_4$, 4-$BrC_6H_4$, 4-$O_2NC_6H_4$, 1-naphthyl, 2-naphthyl, 4-$PhC_6H_4$, 3-PhO-4-$FC_6H_3$, (*E*)-PhCH=CH, 4-$MeOC_6H_4$CH=CH, 2-furyl, 2-thienyl, Cy

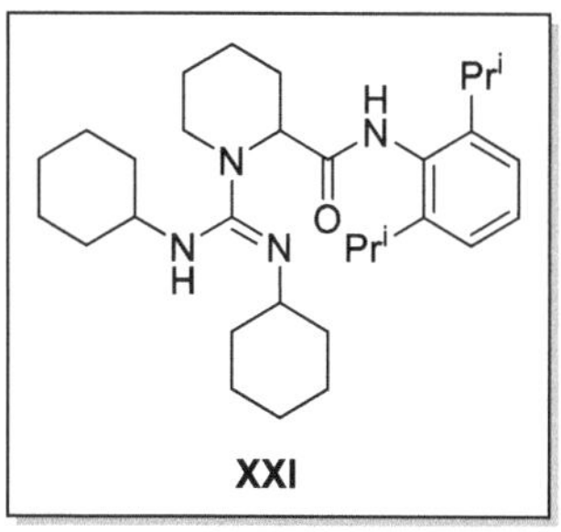


**Scheme 31** Addition of β-keto esters **68** to nitroalkenes

Excellent yields and enantioselectivities as well as diastereoselectivities were obtained in the addition of cyclic β-keto esters **68** to nitroalkenes **64** in the presence of the guanidine **XXI**, the most active among several derivatives with similar structure, giving the expected products **69** [41] (Scheme 31). The same process was also applied to a nitrodiene affording exclusively the corresponding 1,4-adduct with outstanding yield and stereoselection. Based on NMR and X-ray analysis, a dual activation by the organocatalyst was proposed in order to explain the observed stereochemistry in the products **69**.

Dialkyl malonates **2** added efficiently to nitrostyrene (**64**) with high yields and enantioselectivities in the presence of catalytic amounts of the 2-aminobenzimidazole related guanidine **XXII** and trifluoroacetic acid as cocatalyst giving the expected compounds **65** with good results [42] (Scheme 32). The application of the same protocol for α-substituted β-keto esters gave variable diastereoselectivities. One interesting feature of this process is that the guanidine catalyst can be easily recovered by simple acid–base extractive workup. DFT calculations support that the protonated cyclohexylamine moiety activates the nitroalkene, and the benzimidazole unit activates the nucleophile. Thus, an opposite activation mode as compared to that of the corresponding thiourea was observed.

The conjugate addition of dimethyl and diethyl malonates **2** to nitroalkenes **64** has been achieved in an enantioselective manner by using guanidines **XXIII** and

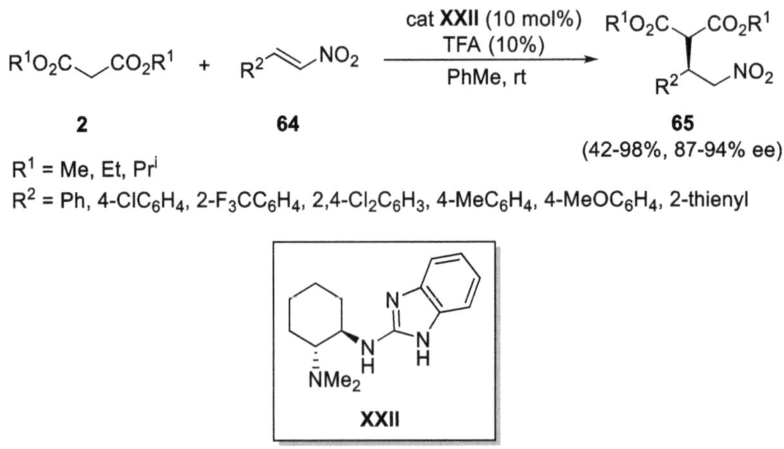


**Scheme 32** Addition of malonates to nitrostyrenes

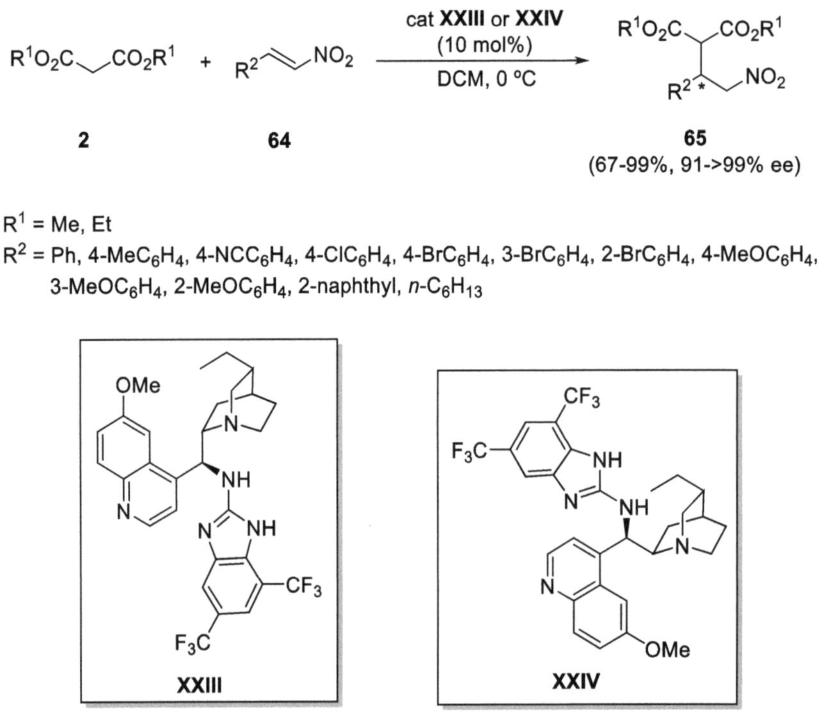


**Scheme 33** Addition of malonates to nitroalkenes

**XXIV** with a 2-aminobenzimidazole structure as catalysts, so that the corresponding enantiomers of type **65** could be isolated with excellent results [43, 44] (Scheme 33). Also in this case, a bifunctional behaviour of the catalyst

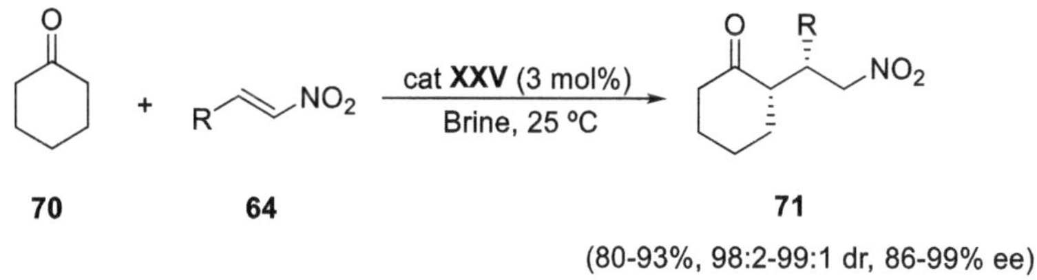


R = Ph, 4-$FC_6H_4$, 4-$ClC_6H_4$, 2-$ClC_6H_4$, 2,4-$Cl_2C_6H_3$, 4-$BrC_6H_4$, 4-$MeC_6H_4$, 4-$MeOC_6H_4$, 3-$MeOC_6H_4$, 3-$MeOC_6H_4$, 3,4-$(MeO)_2C_6H_3$, 2-furyl, Cy

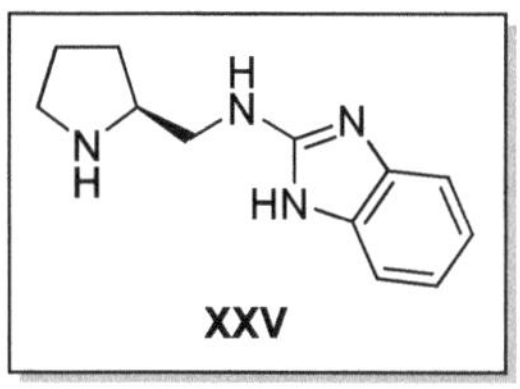


**Scheme 34** Addition of cyclohexanone to nitroalkenes

was proposed in order to explain the results: whereas the two NH groups of the catalyst activate the nitro group of the nitroalkene, the protonated bicyclic amine interacts with the malonate, thus locating both reagents in the correct arrangement to afford the obtained products with the right stereochemistry.

The pyrrolidine derivative **XXV** was a very active organocatalyst for the addition of cyclohexanone (**70**) to nitrostyrene (**64**) yielding the expected products **71** with very good results in terms of yields and diastereo- and enantioselectivities. Whereas cyclopentanone gave also good results, acetone or aliphatic aldehydes did not work properly under the same reaction conditions [45] (Scheme 34). In order to get good results, an aromatic carboxylic acid [$PhCO_2H$, 2-$MeOC_6H_4CO_2H$, 3-$MeOC_6H_4CO_2H$, 4-$MeOC_6H_4CO_2H$, 3,5-$(O_2N)C_6H_3CO_2H$] or *p*-toluenesulfonic acid were used as cocatalysts.

Very moderate results were obtained when guanidines **XXVI** and **XXVII** were used as catalysts for the addition of indole (**72**) to nitrostyrene (**64**) under acidic conditions to yield the corresponding product **73** [46] (Scheme 35). In this study, which included 15 catalysts, the best results were obtained using other types of thioureas without the guanidine moiety.

The enantioselective Michael addition of isobutyraldehyde (**56**, $R^1$=Me) to nitroarenes **64** has been reported in a process catalysed by the chiral guanidine **XVI**, so the expected products **74** were obtained with good results in the presence of imidazole as a cocatalyst [47] (Scheme 36). It is assumed that an enamine of the primary amine of the catalyst and the aldehyde is initially formed. The activation of the nitro group by the guanidine moiety through hydrogen bonds as proposed in the case of maleimides (see Scheme 24) was discarded on the basis of DFT calculations and experimental results. The guanidine group just blocks one of the faces of the enamine, with water molecules activating the nitro group.

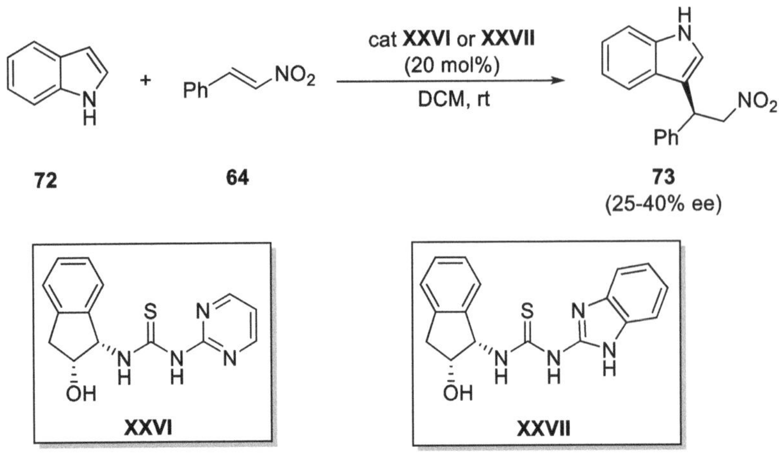


Scheme 35 Addition of indole to nitrostyrene

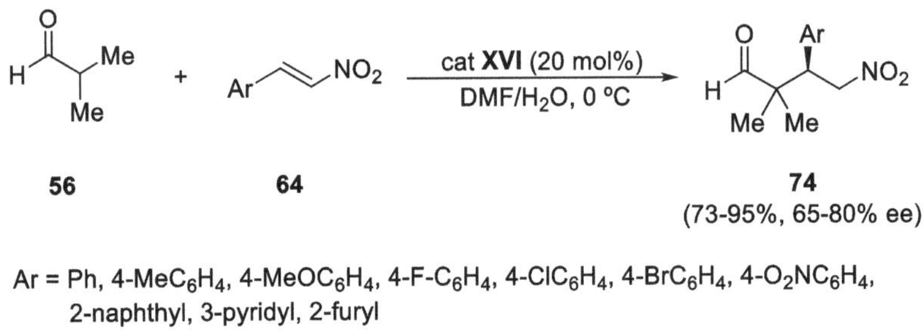


Scheme 36 Addition of isobutyraldehyde to nitroarenes

# 3 Oxygen Nucleophiles

The enantioselective Michael addition of oxygen-containing nucleophiles to electrophilic olefins promoted by a chiral guanidine has been explored for intramolecular reactions. Thus, the sulfonyl guanidine **XX** was used in the cyclization of hydroxylamines **75** and phenol derivatives **76** to afford the corresponding products **77** and **78**, respectively, in an enantioselective manner [48] (Scheme 37). This methodology has been applied to the synthesis of interesting pharmacologically active compounds, such as raxofelast (**79**).

Modest results in terms of yield and enantioselectivity were obtained in the application of the phenol cyclization shown in Scheme 37 to the corresponding six-membered rings. Thus, 2,2-disubstituted chromanes **81** were obtained in low to moderate ee, starting from phenols **80** by using the chiral guanidine **I** as catalyst. A little bit better results were achieved by modifying the structure of the catalyst **I** by changing the substitution at the phenyl group and at the nitrogen atoms [49] (Scheme 38). The *E*/*Z* geometry of the starting material and bulkiness of the R

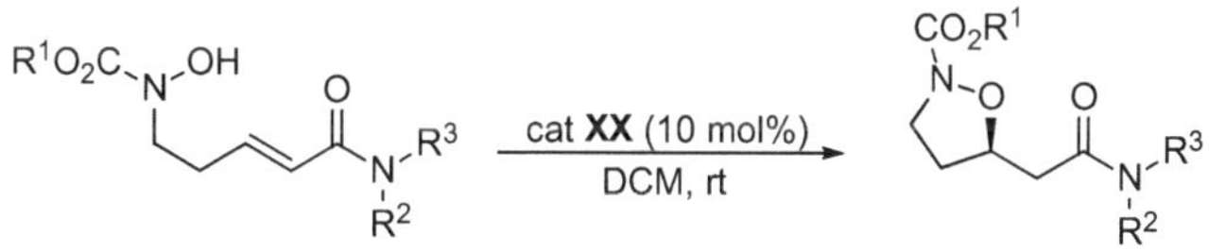


75

77
(70-99%, 84-96% ee)

$R^1$ = Me, Bn
$R^2$, $R^3$ = H, Me, Ph, Bn, $Pr^i$, 4-$MeC_6H_4$, $(CH_2)C_6H_4$

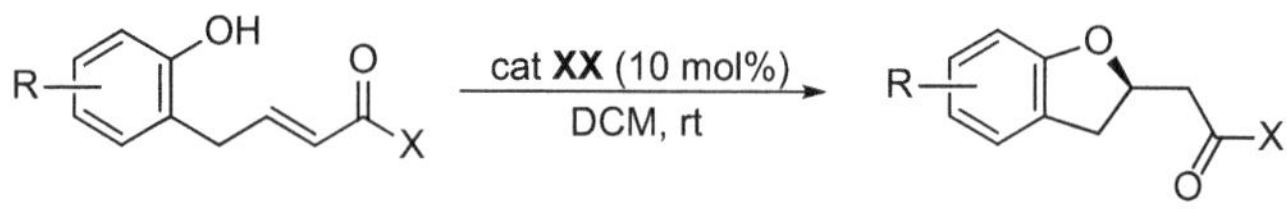


76

78
(41-98%, 83-98% ee)

R = H, 2-Me, 2-MeO, 4-MeO, 4-$NO_2$, 2-Br
X = NMePh, N(OMe)Me, OMe, $OBu^t$, OBn

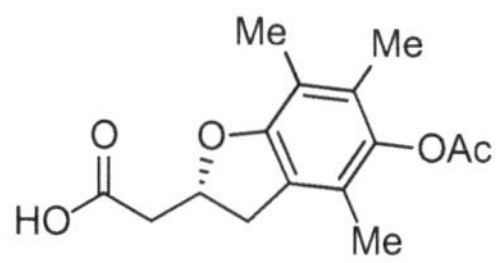

**79**: Raxofelast

**Scheme 37** Cyclization of hydroxylamines **75** and phenols **76**

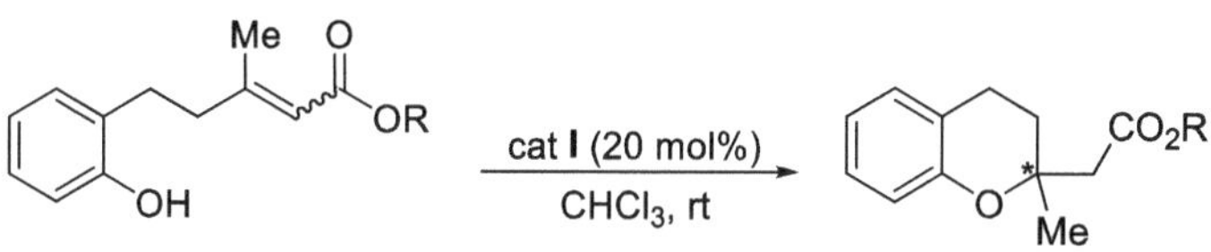


80

81
(7-75%, 6-71% ee)

R = Me, Et, $Bu^t$

**Scheme 38** Cyclization of phenols **80** to chromanes **81**

group played important roles in the asymmetric induction: the best results were obtained using the *Z*-isomer and the less bulky methyl ester.

A special case of conjugate addition of an oxygenated nucleophile to an electron-deficient olefin is its epoxidation [50], which in some cases has been achieved using chiral guanidines. The epoxidation of chalcone derivatives **10** with hydrogen peroxide under basic conditions was realized using the protonated guanidine **XXVIII** (related to the catalyst **XVIII**) giving the corresponding epoxides **82** with excellent yields and enantioselectivities [51] (Scheme 39). The bifunctional organocatalyst contains two urea and one guanidine moieties, which were

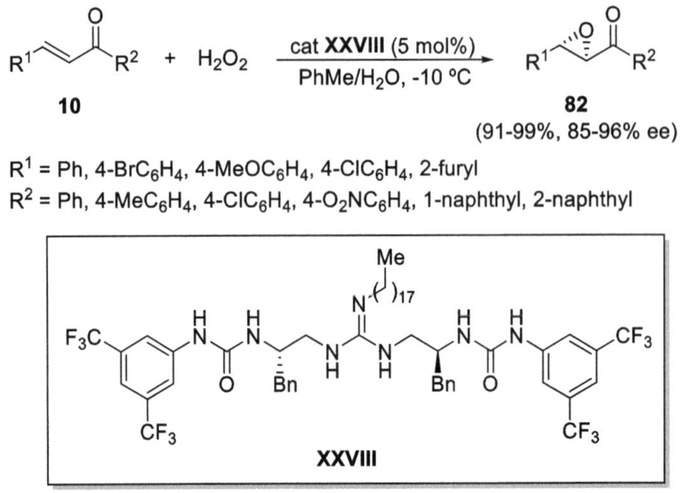


**Scheme 39** Enantioselective epoxidation of chalcones

suggested to act cooperatively by guanidine–hydrogen peroxide and urea–enone interactions, respectively.

Other accounts on the asymmetric epoxidation of chalcone with hydroperoxides or sodium hypochlorite have been reported using guanidines **I** [12], **XXIX** [52], **XXX** (as $HBF_4$ salt) [53] and **XXXI** [54] with very variable results (34–99%, 4–93% ee).

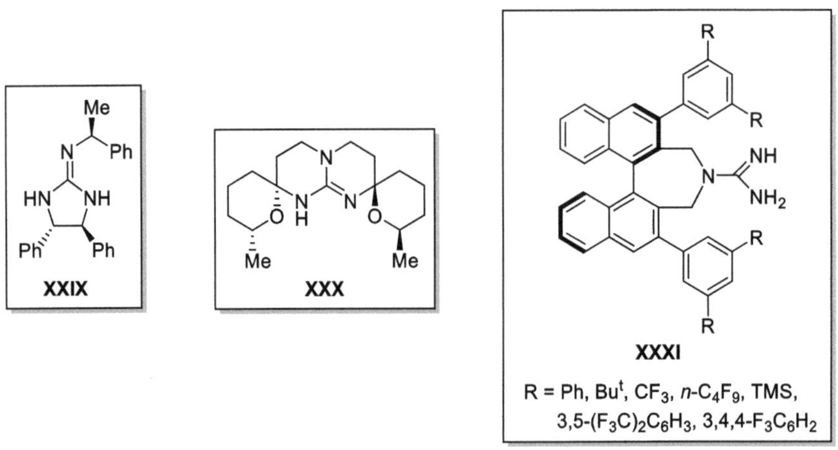


The epoxidation of the cyclic dienone **83** using *tert*-butyl hydroperoxide and guanidines **XXXII** [55] and **XXXIII–XXXV** in stoichiometric amounts [56] gave the corresponding monoepoxide **84** with modest yields and enantioselectivities

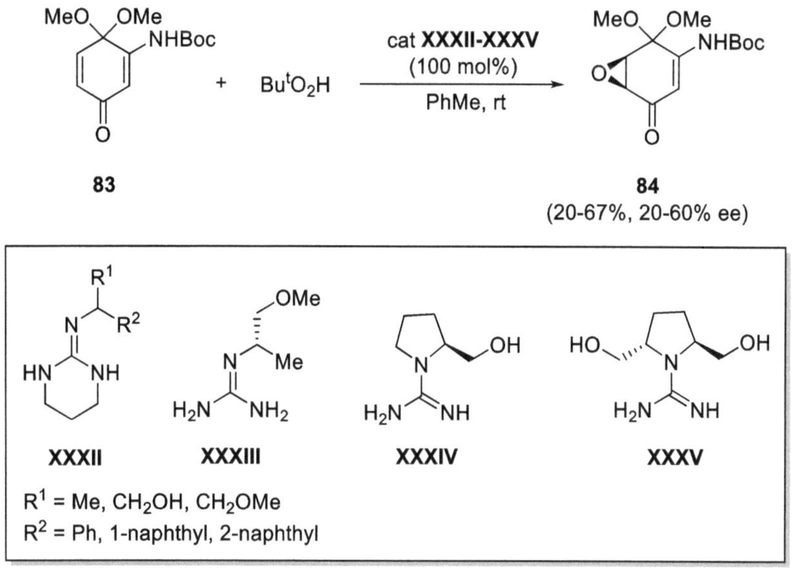


**Scheme 40** Epoxidation of compound **83**

(Scheme 40). The obtained epoxycyclohexenone core is present in several antibiotic and anticancer natural products.

# 4 Nitrogen Nucleophiles

The most studied reaction in this section has been the conjugate addition of pyrrolidine (**85**) to the α,β-unsaturated lactone **48**. A significant acceleration of the mentioned reaction was reported by using a catalytic amount of functionalized guanidines of type **XXXVI** [57, 58] or **XXXVII** [59], but no information about enantioselectivity is given.

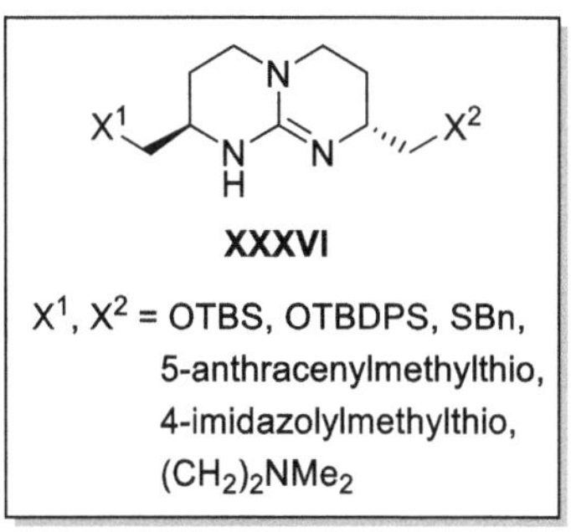


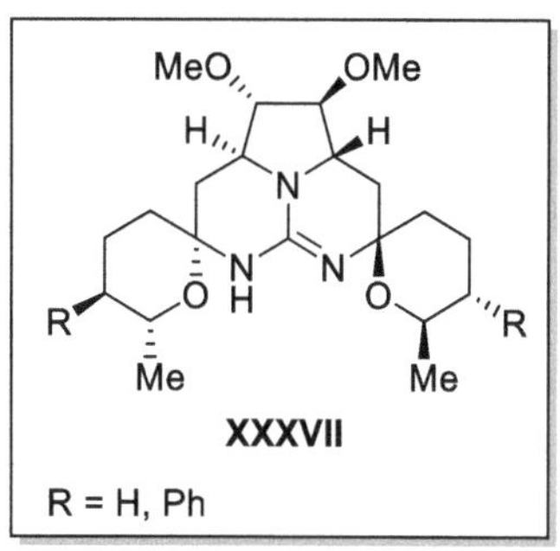

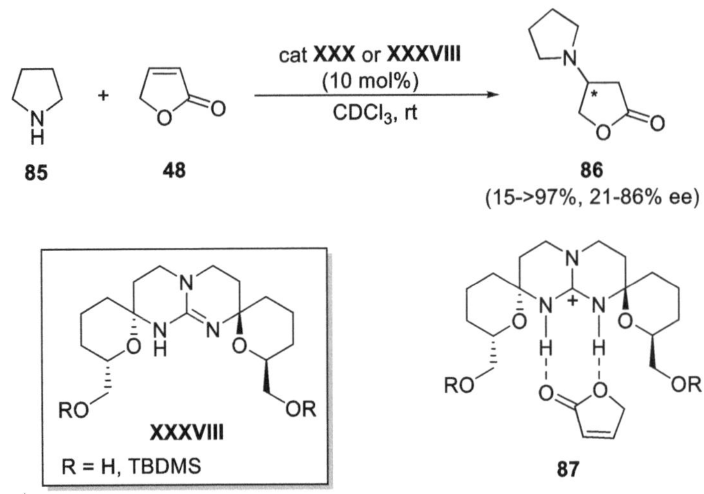


**Scheme 41** Addition of pyrrolidine to the butenolide **48**

The guanidine **XXX** and its functionalized derivatives **XXXVIII** were used as their salts (HCl, $HBF_4$) in the addition of pyrrolidine (**85**) to the butenolide **48** with variable results concerning conversion and enantioselectivity for the product **86** [53] (Scheme 41). An intermediate species, such as **87**, introducing a chiral environment at the electrophilic olefin **48**, was proposed to be involved in the process.

An intramolecular version of the conjugate addition of a nitrogen-containing moiety to a α,β-unsaturated ester in compound **88** led to the formation of the expected products **89** in a process catalysed by the guanidine **XXXIX** as a salt, in general with very good results [60] (Scheme 42). A variant of this process was also performed in the presence of NBS, yielding compounds **90**. Interestingly, *trans*-**90** could be transformed into the corresponding *cis*-isomers by treatment with potassium hydroxide in toluene, a $S_E1$ mechanism having been postulated to yield the most thermodynamically stable product.

## 5 Sulphur Nucleophiles

To the best of our knowledge, there is only one case of the addition of different thiols **91** to the α-phthalimido acrylate **92** in the presence of a catalytic amount of the guanidine **IIId**. Compounds **93**, precursors of cysteine derivatives, were obtained with excellent yields and enantioselectivities [61] (Scheme 43). Similar results were obtained starting from acrylates of type **92** with substituents (Me, F, Cl) at the aromatic ring.

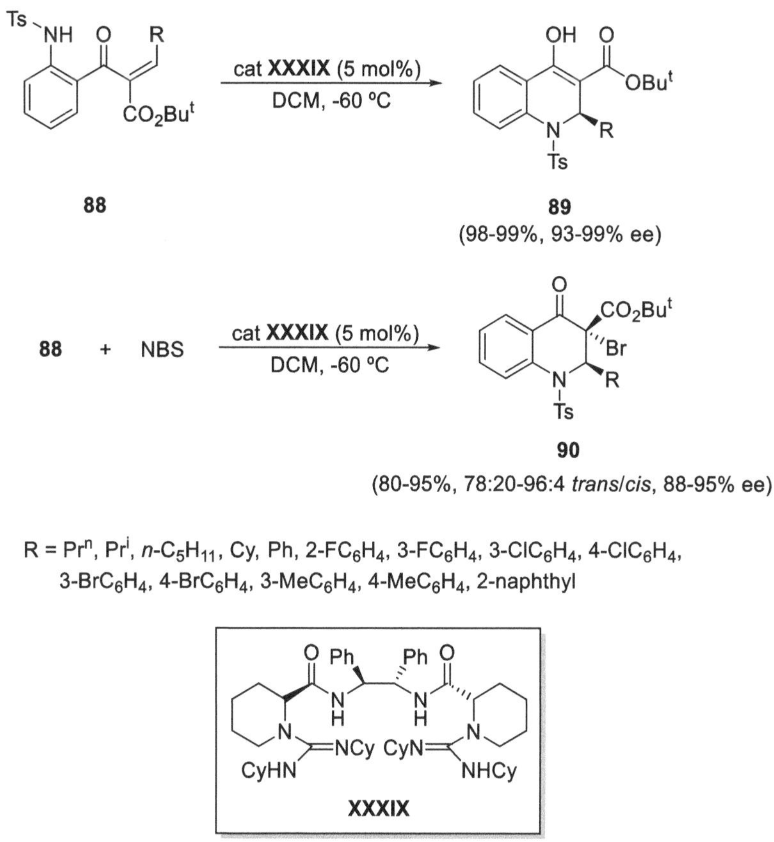


**Scheme 42** Cyclization of compound **88** to **89** and **90**

# 6 Phosphorous Nucleophiles

Phosphine oxides **94** were enantioselectively added to nitrostyrenes **64** in the presence of the guanidine **IIId** to yield the expected products **95** with generally excellent results [62] (Scheme 44). When β-methyl- or β-ethyl-β-nitrostyrene were used with compound **94** having R=1-naphthyl, an excellent diastereoselectivity (95:5) was obtained in favour of the *anti* diastereomer.

The same catalyst **IIId** used in Scheme 44 was also very efficient for the addition of phosphine oxides **94** to the methylene succinimide **96** (*N*-arylitaconamide), so the corresponding products **97** were isolated with excellent results [61] (Scheme 45). The influence of different substituents (alkyl, benzyl, aryl) in compound **96** was investigated, concluding that substituents at the 2- and 6-positions were crucial for the high enantioselectivity, with the highest ee values being obtained with compound **96**. For compounds **94** bearing electron-withdrawing substituents at the aromatic ring (R=aryl), the reaction rates were higher, so lower temperatures were required to obtain high enantioselectivity.

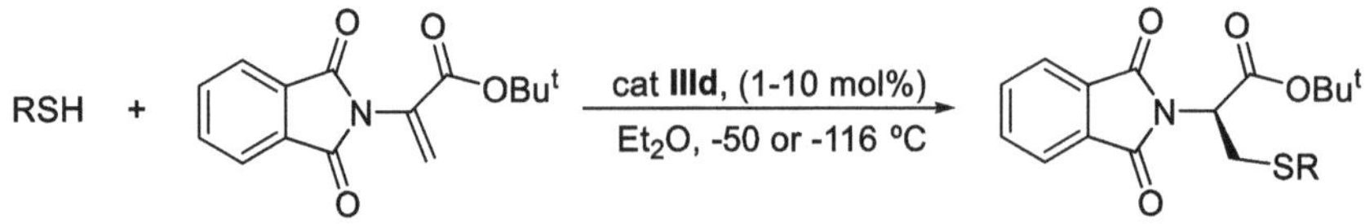

**91** **92** **93** (82-99%, 84-94% ee)

R = Ph, 2-$F_3CC_6H_4$, 2-$MeOC_6H_4$, 4-$BrC_6H_4$, 4-$Bu^tC_6H_4$, 3,5-$Me_2C_6H_3$, 2-thienyl, 1-naphthyl, 4-$HOC_6H_4$, 4-$H_2NC_6H_4$

**Scheme 43** Addition of thiols **91** to the acrylate **92**

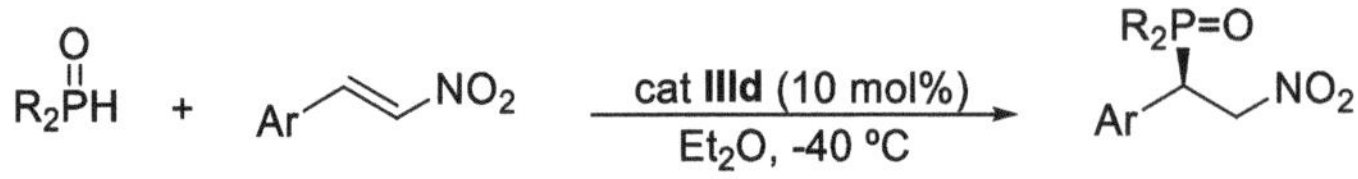

**94** **64** **95** (64-99%, 50-96% ee)

R = Ph, 4-$FC_6H_4$, 4-$PhC_6H_4$, 2-$EtC_6H_4$, 1-naphthyl, 2-naphthyl
Ar = Ph, 2-$ClC_6H_4$, 3-$ClC_6H_4$, 4-$ClC_6H_4$, 4-$BrC_6H_4$, 2-$O_2NC_6H_4$, 3-$O_2NC_6H_4$, 4-$MeC_6H_4$, 2-naphthyl

**Scheme 44** Addition of phosphine oxides to nitrostyrenes

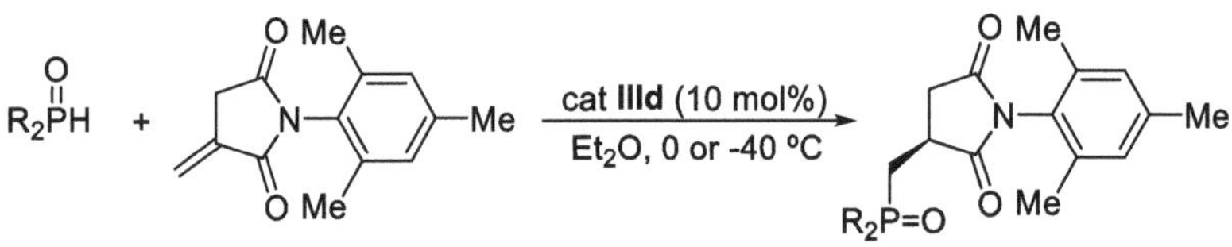

**94** **96** **97** (88-95%, 87-98% ee)

R = Ph, 2-$EtC_6H_4$, 2-$FC_6H_4$, 3-$ClC_6H_4$, 2-$F_3CC_6H_4$, 3-$F_3CC_6H_4$, 3-$FC_6H_4$, 1-naphthyl, 2-methyl-1-naphthyl

**Scheme 45** Addition of phosphine oxides to the succinimide **96**

The guanidine **XL** (of type **XVII**) was shown to be an excellent catalyst for the enantioselective addition of diphenyl phosphite **98** to aliphatic and aromatic nitroalkenes **64** to afford the corresponding products **99** with very good results [63] (Scheme 46). The reduction of compounds **99** with the $NiCl_2$–$NaBH_4$ combination afforded the corresponding amino derivatives preserving the stereochemical integrity. The resulting β-amino phosphonic esters are an attractive class of compounds owing to their potent biological activities as non-proteinogenic analogues of β-amino esters.

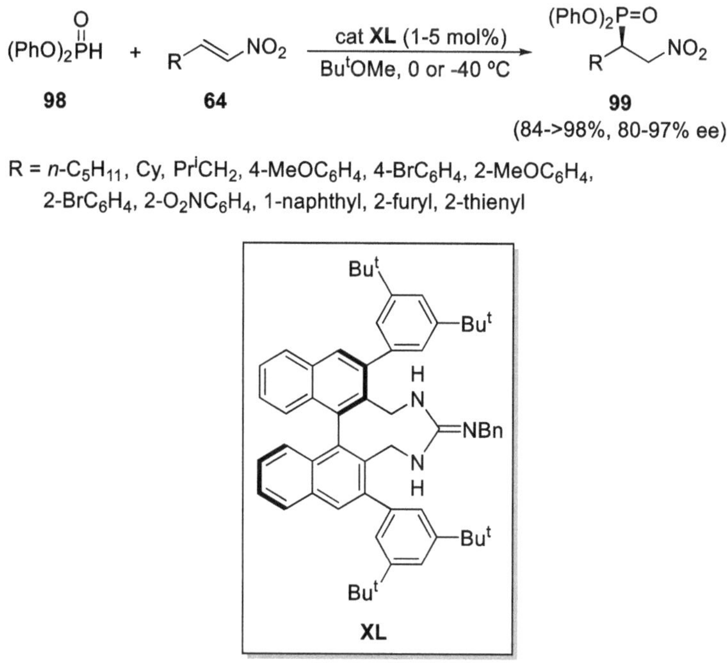


**Scheme 46** Addition of diphenyl phosphite to nitroalkenes

# 7 Cyclizations

In some cases, depending on the structure of the reagents, the conjugate addition can give a cyclization process in a second step through a tandem reaction. This is the case in the reaction of the unsaturated aldehyde **100** with chalcone (**10**) in the presence of the guanidine **XLI** and a base. **XLI** is a precursor of the corresponding *N*-heterocyclic carbene, the real catalyst in the process. Using this methodology, the cyclopentene derivative **101** is obtained with modest results [64] (Scheme 47). The role of the in situ generated carbene is to form the chiral conjugated Breslow intermediate derived from the aldehyde **100. XLI** thus acts as a bifunctional organocatalyst due to the presence of the additional guanidine moiety activating the electrophilic olefin **10** through hydrogen bonding.

The guanidine **XX** is able to isomerize the acetylenic ester **102** to the corresponding allenyl derivative, which then undergoes a tandem Michael addition–1,3-dipolar cycloaddition with the imine **103**, yielding the substituted pyrrolidine **104** in an enantio- and diastereoselective manner [40] (Scheme 48).

A direct 1,3-dipolar cycloaddition was easily achieved between imines **105** and dimethyl maleate (**106**) using the guanidine **XLII** as the promoter, affording the corresponding *endo*-pyrrolidines **107** with variable yields and, in general, good enantioselection [65] (Scheme 49). Experimental evidence suggests that the

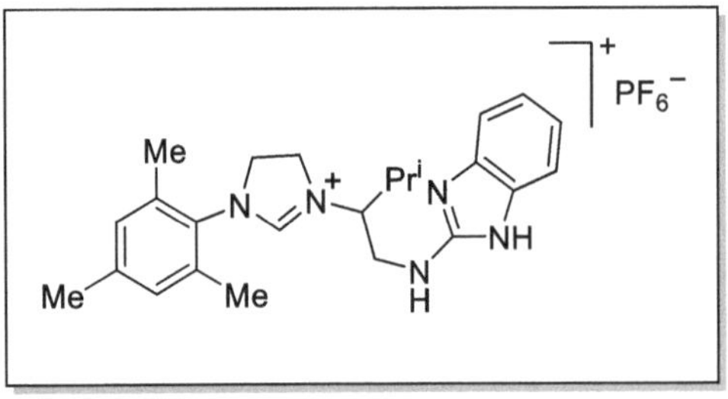


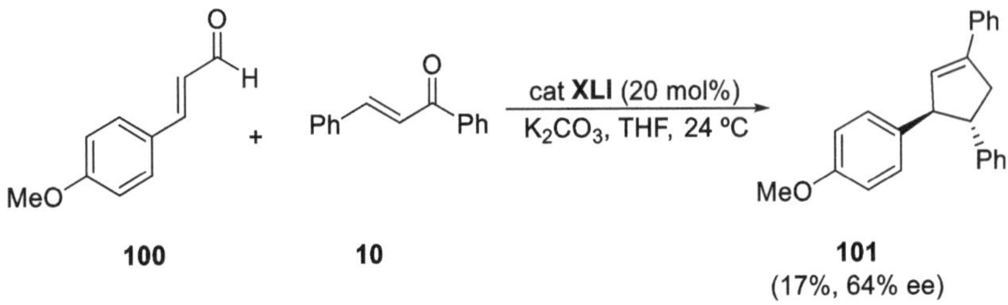


**Scheme 47** Preparation of the cyclopentene **101**

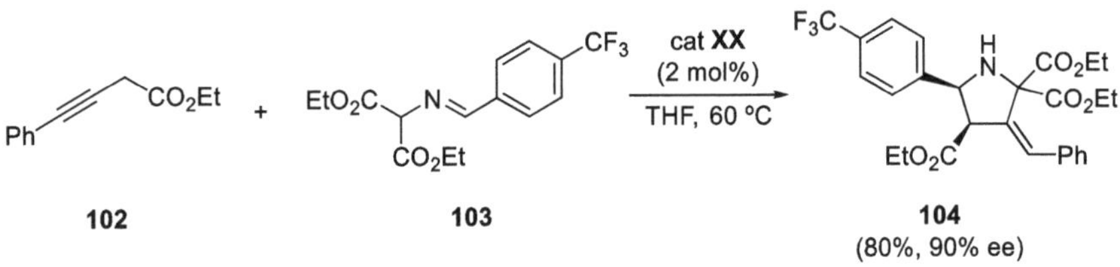


**Scheme 48** Preparation of the pyrrolidine **104**

cycloaddition products were obtained through a concerted pathway: it is considered that the acid–base dual function of the axially chiral guanidine through the double hydrogen-bonding interaction is crucial to accelerate the present [3+2] cycloaddition reaction in comparison with catalysis by achiral organic bases.

A tandem Michael addition–cyclization from anthrones **54** and maleimides **52** in the presence of the guanidine **IIIa** allowed the preparation of the bicyclic compounds **108** with excellent yields and enantioselectivities [31] (Scheme 50). It is worthy to note that the catalyst **IIIa** can tolerate a range of substituents and substitution pattern on the anthrone **54**.

The guanidine derivative **XXXIX** was successfully used for the hetero-Diels–Alder reaction between chalcones **10** and azlactones **35**, which starts from a Michael addition followed by a cyclization. Compounds **109** were obtained with good yields and excellent enantioselectivities [66] (Scheme 51). On the basis of experimental data, a bifunctional activated transition state for the inverse-electron-demand Diels–Alder reaction is proposed: whereas one NH group of the amide acts as a Brønsted acid activating the chalcone by lowering its LUMO energy, the other

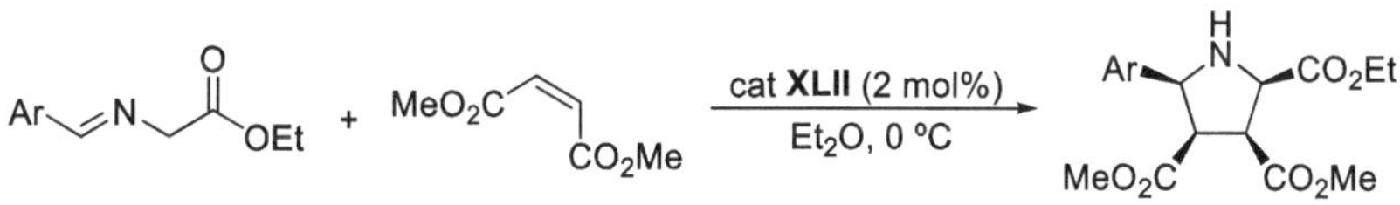


**105** **106** **107**

(21-80%, >98:2 dr, 59-91% ee)

Ar = Ph, 3-$MeOC_6H_4$, 4-$MeOC_6H_4$, 4-$F_3CC_6H_4$, 2-$BrC_6H_4$, 3-$BrC_6H_4$, 4-$BrC_6H_4$, 4-$ClC_6H_4$, 4-$MeOCOC_6H_4$

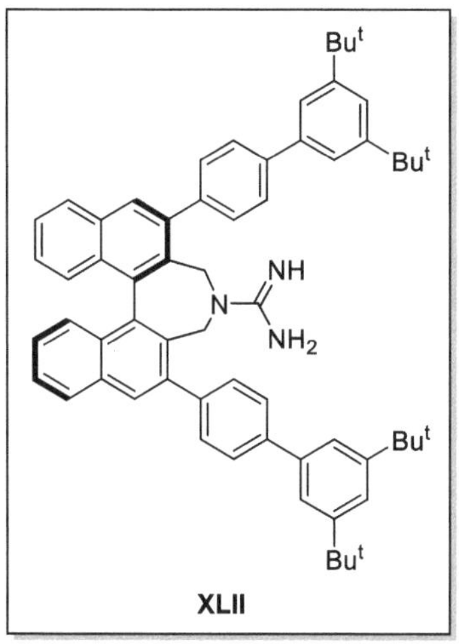


**Scheme 49** 1,3-Dipolar cycloaddition to give prolinates **107**

NH recognizes the enolized azlactone, so it only attacks the *Re*-face of the chalcone, giving the isolated products.

## 8 Final Remarks

In this chapter, the use of several chiral guanidines as organocatalysts in conjugate additions of carbon and heteroatom nucleophiles to electrophilic olefins has been considered. The aim of this chapter has been to present interesting reactions from a synthetic point of view in the frame of enantioselective processes with preparative applications. Mechanistic considerations are scarcely included in this account, but for readers interested in these aspects, the references given can detail them in many cases. Another aspect that has not been considered at all along this chapter is the preparation (for not commercially available compounds) of the chiral guanidines used in this chemistry: also in this case, such preparation can be found in the original literature cited at the end of the chapter.

**54** **52** **108**

(85-96%, 85-99% ee)

$R^1$, $R^2$, $R^3$, $R^4$ = H, Cl, MeHN

$R^5$ = Et, $Bu^i$, $Bu^t$, Cy, Ph, Bn, 2-$O_2NC_6H_4$, 2,5-$Cl_2C_6H_3$, 4-$ClC_6H_4$, 2,6-$F_2C_6H_3$, 2-$MeOC_6H_4$, 2,4,6-$Me_3C_6H_2$

**Scheme 50** Preparation of bicyclic derivatives **108**

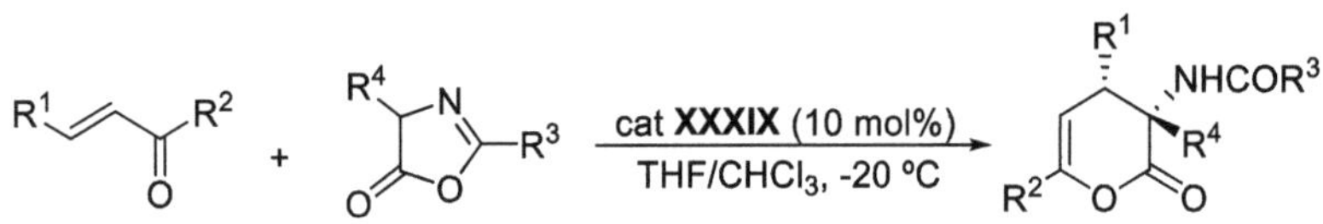


**10** **35** **109**

(40-88%, 89-99% ee)

$R^1$ = Ph, 2-$ClC_6H_4$, 3-$ClC_6H_4$, 4-$ClC_6H_4$, 3,4-$Cl_2C_6H_3$, 4-$FC_6H_4$, 3-$BrC_6H_4$, 4-$BrC_6H_4$, 3-$MeC_6H_4$, 4-$MeC_6H_4$, 3-$PhOC_6H_4$, 4-$PhC_6H_4$, 1-naphthyl, 2-naphthyl, 4-pyridyl, 3,4-$(OCH_2O)C_6H_3$

$R^2$ = Ph, 4-$FC_6H_4$, 4-$BrC_6H_4$, 3-$MeC_6H_4$, 4-$MeC_6H_4$, 2-naphthyl, cinnamyl, 2-furyl, 2-thienyl, 3,4$(OCH_2O)C_6H_3$

$R^3$ = Ph, 3-$ClC_6H_4$, 4-$ClC_6H_4$, 3-$MeC_6H_4$, 3-$MeOC_6H_4$

$R^4$ = Me, Bn, $MeS(CH_2)_2$

**Scheme 51** Hetero Diels–Alder reaction to give lactones **109**

As a conclusion, the methodology presented here (guanidine-catalysed enantioselective Michael additions) is a good choice for the preparation of functionalized chiral compounds in an enantioselective manner.

# References


1. Ishikawa T, Kumamoto T (2006) Synthesis 737
2. Ishikawa T (2009) In: Ishikawa T (ed) Superbases for organic synthesis. Wiley, Chichester, p 93
3. Dalko PI (2007) Enantioselective organocatalysis. Wiley-VCH, Weinheim
4. List B (2009) Asymmetric organocatalysis. Springer, Berlin
5. Pellissier H (2012) Recent developments in asymmetric organocatalysis. RSC, Cambridge
6. Almasi D, Alonso D (2007) Tetrahedron Asymmetry 18:29
7. Leow D, Tan C-H (2009) Chem Asian J 4:488

8. Nagasawa K, Sohtome Y (2012) In: Maruoka K (ed) Asymmetric organocatalysis 2. Brönsted base and acid catalysis and additional topics. Georg Thieme, Stuttgart
9. Selig P (2013) Synthesis 703
10. Chauban P, Kaur J, Chimi SS (2003) Chem Asian J 8:328
11. Vicario JL, Badía D, Carrillo L, Reyes E (2010) Organocatalytic enantioselective conjugate addition reactions. A powerful tool for the stereocontrolled synthesis of complex molecules. RSC, Cambridge
12. Kumamoto T, Ebine K, Endo M, Araki Y, Fushimi Y, Miyamoto I, Ishikawa T, Iosbe T, Fukuda K (2005) Heterocycles 66:347
13. Pan QB, Ma DW (2003) Chin J Chem 21:793
14. Kaneko S, Yoshino T, Katoh T, Terashima S (1998) Tetrahedron 54:5471
15. Ye W, Leow D, Goh SLM, Tan C-T, Chian C-H (2006) Tetrahedron Lett 47:1007
16. Ye W, Jiang Z, Xhao Y, Goh SLM, Leow D, Soh Y-T, Tan C-H (2007) Adv Synth Catal 349: 2454
17. Jiang Z, Yang Y, Pan Y, Zhao Y, Liu H, Tan C-H (2009) Chem Eur J 15:4925
18. Jiang Z, Pan Y, Zhao Y, Ma T, Lee R, Yang Y, Huang K-W, Wong MW, Tan C-H (2009) Angew Chem Int Ed 48:3627
19. Jiang Z, Ye W, Yang Y, Tan C-H (2008) Adv Synth Catal 350:2345
20. Davis AP, Dempsey KJ (1995) Tetrahedron Asymmetry 6:2829
21. Linton BR, Reutershan MH, Aderman CM, Richardson EA, Browell KR, Ashley CW, Evans CA, Miller SJ (2007) Tetrahedron Lett 48:1993
22. Tsogoeva SB, JagTap SB, Ardemasova ZA, Kalikhevich VN (2004) Eur J Org Chem 4014
23. Dai Q, Huang H, Zhao CG (2013) J Org Chem 78:4153
24. Ma D, Cheng K (1999) Tetrahedron Asymmetry 10:713
25. Misaki T, Kawano K, Sugimura T (2011) J Am Chem Soc 133:5695
26. Yang Y, Dong S, Lin X, Lin L, Feng X (2012) Chem Commun 48:5040
27. Ryoda A, Yajima N, Haga T, Kumamoto T, Nakanishi W, Kawahata M, Yamaguchi K, Ishikawa T (2008) J Org Chem 73:130
28. Gómez-Torres E, Alonso DA, Gómez-Bengoa E, Nájera C (2011) Org Lett 13:6106
29. Gómez-Torres E, Alonso DA, Gómez-Bengoa E, Nájera C (2013) Eur J Org Chem 1434
30. Peng B, Cheng K-J, Ma D-W (2000) Chin J Chem 18:411
31. Shen J, Nguyen TT, Goh Y-P, Ye W, Fu X, Xu J, Tan C-H (2006) J Am Chem Soc 128:13692
32. Avila A, Chinchilla R, Nájera C (2012) Tetrahedron Asymmetry 23:1625
33. Avila A, Chinchilla R, Gómez-Bengoa E, Nájera C (2013) Eur J Org Chem 5085
34. Li L, Chen W, Yang W, Pan Y, Liu H, Tan C-H, Jiang Z (2012) Chem Commun 48:5124
35. Ishikawa T, Araki Y, Kumamoto T, Seki H, Fukuda K, Isobe T (2001) Chem Commun 245
36. Terada M, Ube H, Yaguchi Y (2006) J Am Chem Soc 128:1454
37. Terada M (2010) J Synth Org Chem Jpn 68:1159
38. Sohtome Y, Shin B, Horitsugi N, Takagi R, Noguchi K, Nagasawa K (2010) Angew Chem Int Ed 49:7299
39. Sohtome Y, Yamaguchi T, Shin B, Nagasawa K (2011) Chem Lett 40:843
40. Inokuma T, Furukawa M, Uno T, Suzuki Y, Yoshida K, Yano Y, Matsuzaki K, Takemoto Y (2011) Chem Eur J 17:10470
41. Yu X, Liu X, Lin L, Feng X (2009) Angew Chem Int Ed 48:5195
42. Almasi D, Alonso DA, Gómez-Bengoa E, Nájera C (2009) J Org Chem 74:6163
43. Zhang L, Lee M-M, Lee S-M, Lee J, Cheng M, Jeong B-S, Park H-G, Jew S-S (2009) Adv Synth Catal 351:3063
44. Lee M, Zhang L, Park Y, Park H-G (2012) Tetrahedron 68:1452
45. Lin J, Tian H, Jiang Y-J, Huang W-B, Zheng L-Y, Zhang S-Q (2011) Tetrahedron Asymmetry 22:1434
46. Ganesh M, Seidel D (2008) J Am Chem Soc 130:16464
47. Avila A, Chinchilla R, Fiser B, Gómez-Bengoa E, Nájera C (2014) Tetrahedron Asymmetry 25:462

48. Kobayashi Y, Taniguchi Y, Hayama N, Inokuma T, Takemoto Y (2013) Angew Chem Int Ed 52:1114
49. Saito N, Ryoda A, Nakanishi W, Kumamoto T, Ishikawa T (2008) Eur J Org Chem 2759
50. Weiβ KM, Tsogoeva SB (2011) Chem Rec 11:18
51. Tanaka S, Nagasawa K (2009) Synlett 667
52. Ichikawa T, Isobe T (2002) Chem Eur J 8:552
53. Allinham MT, Howard-Jones A, Murphy PJ, Thomas DA, Caulkett PWR (2003) Tetrahedron Lett 44:8677
54. Terada M, Ube H, Yokohama S, Shimizu H (2008) US Patent 2008/0154036-A1
55. McManus JC, Genski T, Carey JS, Taylor RJK (2003) Synlett 369
56. McManus JC, Carey JS, Taylor RJK (2003) Synlett 365
57. Alcázar V, Morán JR, de Mendoza J (1995) Tetrahedron Lett 36:3941
58. Martín-Portugués M, Alcázar V, Prados P, de Mendoza J (2002) Tetrahedron 58:2951
59. Nagasawa K, Georgieva A, Takahashi H, Nakata T (2001) Tetrahedron 57:8959
60. Xiao X, Liu X, Dong S, Cai Y, Lin L, Feng X (2012) Chem Eur J 18:15922
61. Leow D, Lin S, Chittimalla SK, Fu X, Tan C-H (2008) Angew Chem Int Ed 47:5641
62. Fu X, Jiang Z, Tan C-H (2007) Chem Commun 5058
63. Terada M, Ikehara T, Ube H (2007) J Am Chem Soc 129:14112
64. Nawaz F, Zaghouani M, Bonne D, Chuzel O, Rodriguez J, Coquerel Y (2013) Eur J Org Chem 8253
65. Nakano M, Terada M (2009) Synlett 1670
66. Dong S, Liu X, Chen X, Mei F, Zhang Y, Gao B, Lin L, Feng X (2010) J Am Chem Soc 132:10650

Top Heterocycl Chem (2017) 50: 129–156
DOI: 10.1007/7081_2015_175

Published online: 29 August 2015

# Chiral Bicyclic Guanidine, Bis-Guanidinium, Pentanidium and Related Organocatalysts

**Dasheng Leow and Choon-Hong Tan**

**Abstract** The roles of chiral bicyclic guanidine, bis-guanidinium and pentanidium in catalysis are presented. Besides the usual Brønsted base reactivity, less known modes of interactions of these catalysts, such as Brønsted base–Brønsted acid and Brønsted–Lewis acid bifunctional activations will be discussed in this account. Pentanidiums are 'super-guanidines' with extended conjugation, and their roles in catalysis and halogen bonding will also be elaborated.

**Keywords** Bicyclic guanidine, Bis-guanidinium, Brønsted base, Organocatalysis, Pentanidium

## Contents



D. Leow (✉)
Department of Chemistry, National Tsing Hua University, 101, Sec 2, Kuang-Fu Rd, Hsinchu 30013, Taiwan, Republic of China
e-mail: dsleow@mx.nthu.edu.tw

C.-H. Tan
Division of Chemistry and Biological Chemistry, Nanyang Technological University, 21 Nanyang Link, Singapore 637371, Singapore
e-mail: choonhong@ntu.edu.sg

## Abbreviations

| | |
|---|---|
| BARF | $[B[3,5\text{-}(CF_3)_2C_6H_3]_4]^-$; Tetrakis(3,5-bis(trifluoromethyl)phenyl) borate |
| DBU | 1,8-Diazabicycloundec-7-ene |
| Eoc | *N*-3-Ethylpentan-3-yloxycarbonyl |
| NHPI | *N*-Hydroxyphthalimide |
| TBD | 1,5,7-Triazabicyclo[4.4.0]dec-5-ene |
| TBO | 1,4,6-Triazabicyclo[3.3.0]oct-4-ene |

# 1 Introduction

Guanidines are well known for their roles as superbases (for reviews on organic Brønsted bases, see [1, 2]). However, the utilisation of guanidines as catalysts was less explored until the last decade. Dual hydrogen bonding in chiral urea and thiourea catalysts was first recognised to play an important role in electrophilic activation (for reviews on asymmetric catalysis utilising hydrogen bonding, see [3–5]). In a similar manner, their unique dual hydrogen bonding in the protonated form and their strong basicity enable guanidines to be effective chiral Brønsted base catalysts.

Corey and co-workers first found an enantioselective Strecker reaction using a bicyclic guanidine **1a** (Fig. 1) [6, 7]. Subsequently, we developed the *tert*-butyl bicyclic guanidine **1b** and demonstrated a series of enantioselective transformations. Besides bicyclic guanidines, we also developed non-$C_2$-symmetric, guanidine catalyst **2**, based on an aminoindanol core (Fig. 1).

The p$K_a$ of catalysts **1** and **2** is in the range of 10–12 (ACD Lab, SciFinder), and thus the ranges of protons that can be abstracted by these catalysts are limited by their inherent basicity. If there is an additional guanidine within the same molecule, the two guanidines may have synergistic effect. With this hypothesis, we explored various bis-guanidines Brønsted base catalysts **3** (Fig. 1). Unfortunately, we did not see an increase in basicity. However, we found that Brønsted acid forms of the bis-guanidines demonstrate interesting catalytic activity.

We also hypothesise that when the number of nitrogen atoms in conjugation in the guanidine is increased from 3 to 5, it should form a more basic 'super-guanidine'. The IUPAC name for this 'super-guanidine' is diaminomethylidene guanidine, which is also known as biguanide. To avoid confusion with guanidine or bis-guanidine, we coined a new name for it, calling it pentanidine (base form) and pentanidium (salt form). The corresponding alkylated salt, pentanidium salt **4**, was found to be an excellent phase-transfer catalyst (Fig. 1).

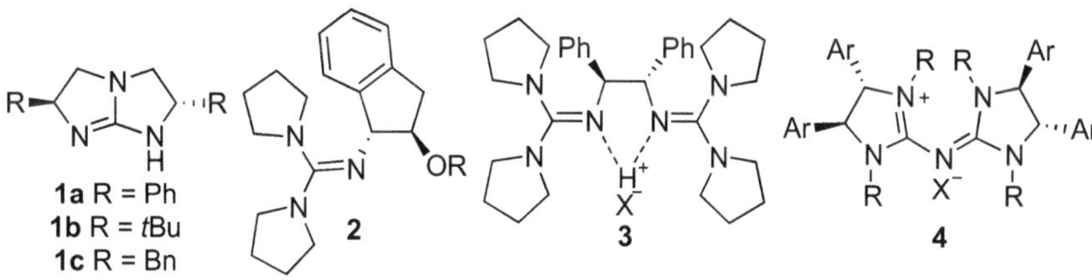


**Fig. 1** Chiral bicyclic guanidines, bis-guanidinium and pentanidium catalysts

This chapter does not aim to cover a broad and comprehensive survey of different chiral guanidine-type organocatalysts. Many excellent reviews have already served this function (for selected reviews on asymmetric guanidine catalysis, see [8–15]). Rather, this chapter reflects the development of guanidine catalysts **1b–c** and **2–4**; thus pointing to possible emerging trends in guanidine research [9, 12]. Hopefully, we will be able to let readers gain new perspectives from this review.

# 2 Chiral Bicyclic Guanidines as Brønsted Base Catalysts

## *2.1 Early Developments*

In 1989, de Mendoza and co-workers reported the extraction of racemic tryptophan using [6,6]-fused ring guanidines **5** as chiral receptors (Fig. 2) [16, 17]. In the same year, Corey and co-workers achieved the synthesis of the more rigid chiral [5,5]-bicyclic guanidine **1a**. However, its catalytic application was not demonstrated until much later [6]. Subsequently, Schmidtchen prepared guanidine **6** to investigate its selective anion binding ability (Fig. 2) [18]. The first example using a chiral bicyclic guanidine as Brønsted base catalyst was demonstrated by Davis, using guanidine **7**, which was shown to catalyse the conjugate addition of nitroalkanes to methyl vinyl ketone (Fig. 2) [19]. However, the observed levels of enantioselectivity were moderate. Another early example was Murphy's bicyclic guanidine **8**, which catalysed the Henry reaction of nitromethane to isovaleraldehyde (Fig. 2) [20, 21]. The enantioselectivities observed were also moderate. Finally, Corey and co-workers reported the asymmetric Strecker reaction of *N*-benzhydryl benzaldimine derivatives using bicyclic guanidine **1a** as catalyst, attaining useful levels of enantioselectivity [7]. [5,5]-Bicyclic guanidines **1** are more rigid and can hold the chiral groups firmly in place, while [6,6]-bicyclic guanidines are both more basic and more nucleophilic.

One critical factor for the successful development of chiral bicyclic guanidine **1b** was the efficient preparation of the catalysts via the aziridine-based synthetic strategy (Scheme 1) [22]. We were able to achieve moderate to good yields for a range of bicyclic guanidine catalysts with different appendages, using a simple and

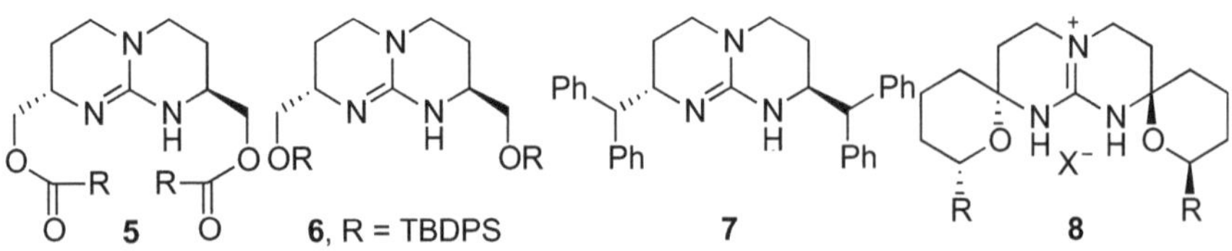


**Fig. 2** Early examples of chiral bicyclic guanidine catalysts

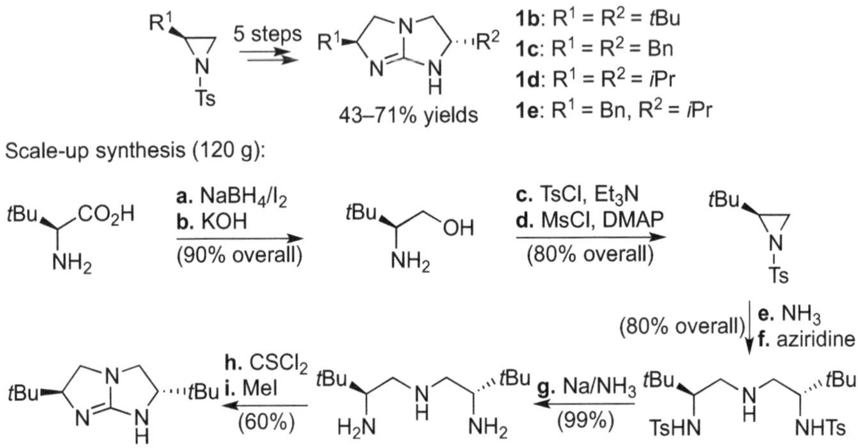


**Scheme 1** Efficient synthesis of chiral bicyclic guanidines. Scale-up synthesis (120 g of amino acid); reagents and conditions: (**a**) $NaBH_4/I_2$, THF, 0°C, then reflux, 18 h; (**b**) MeOH, 0°C, then 20% KOH, 4 h (90% overall); (**c**) TsCl, $Et_3N$, MeCN, 0°C, 20 min; (**d**) MsCl, DMAP, $Et_3N$, DCM, RT (80% overall); (**e**) $NH_3$, MeOH, 0°C to RT, overnight; (**f**) aziridine, MeCN, reflux, 3 d, recrystallised (80% overall); (**g**) Na in liquid $NH_3$, −78°C, 4 h (99%); (**h**) $CSCl_2$, $Et_3N$, $MeNO_2$, 110°C, 2 h; (**i**) MeI, AcOH, reflux, 3 h, then overnight, RT (60%)

efficient five-step synthetic route. We have also demonstrated the scalability of our synthetic procedure for preparing bicyclic guanidine **1b** on a synthetically useful scale (120 g of the amino acid) (for selected reviews on enantioselective protonation, see [23]).

We first investigated the conjugate addition reaction of various carbonyl compounds with electron deficient alkenes using commercially available achiral guanidine **9** as the catalyst (Scheme 2) [24]. Generally, we found that the reactivity of TBD (1,5,7-triazabicyclo[4.4.0]dec-5-ene) **9** was higher than that of other organic bases such as DBU or Hünig's base, due to its higher basicity. When we first attempted to investigate the enantioselective conjugate addition using guanidine **1b**, we met with some difficulties, such as long reaction times and moderate enantioselectivities. We found that the catalytic activities of guanidine **1b** were significantly lower as compared to TBD **9**.

McKay and Kreling found that the $pK_a$ of 1,4,6-triazabicyclo[3.3.0]oct-4-ene (TBO, [5,5]-fused ring) was 10.6 in $H_2O$ versus 11.22 for guanidine **9** (TBD, [6,6]-

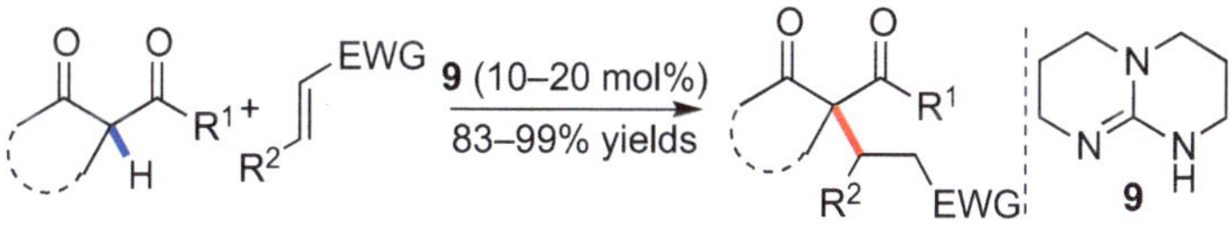


**Scheme 2** Guanidine **9**-catalysed conjugate addition reactions

fused ring) [25]. They attributed the lower p$K_a$ to the considerable strain existing in the 5-membered fused rings. Thus, the resonance stabilisation of the guanidinium ion could not reach the maximum. Recently, Waymouth and Hedrick performed a comparative study between the reactivity profiles of TBO and TBD **9**, using both theoretical and mechanistic approaches [26]. They concluded that the higher basicity and nucleophilicity of TBD **9** increased the reaction rates in general base reactions. We speculated that there might be two further reasons: (a) hydrogen bonding activation provided by guanidine **1b** was not as effective as by TBD **9** due to the deviation of angle for the hydrogen bonds, and (b) the bulky appendages weakened the effects of hydrogen bonding activation.

## 2.2 *Novel Donors*

Due to the lower catalytic activity of guanidine **1b**, the reaction times of Michael reactions of malonic esters with various electron-withdrawing alkenes were longer than desired. Therefore, we adopted a new type of donor for these reactions, the 1,3-dithiomalonates (Scheme 3) [27]. As the overlap of *C*(2p) and *S*(3p) orbitals are poor, the α-proton of 1,3-thioester is more acidic than that of its corresponding ester by about 3 units of p$K_a$. This enhanced the reaction rates and yields significantly. By installing bulky alkyl thioester groups, optimal enantioselectivities were achieved with different alkene substrates, such as maleimides, cyclic enones, furanone and acyclic acceptors. The Michael addition to acyclic 4-oxo-4-arylbut-2-enoate was highly regioselective, occurring only at the β-position of the enone moiety. The main disadvantage of using 1,3-thioesters is the requirement to prepare them, as they are not commercially available.

We found that we can also accelerate the reaction rates of the Michael reactions dramatically using triethylamine as the solvent (Scheme 4) [28]. Useful substrates such as malonic esters and 1,3-diketones were applicable using this condition. We postulated that the enhancement of the reaction rate was due to the stabilisation of the enolate–guanidinium complex by triethylamine.

The guanidine-catalysed Michael reaction was further probed using computational studies [29]. The optimised geometry of protonated guanidine **1b** was found to have a $C_1$ symmetry with distinct top and bottom faces (Fig. 3). The key initial step for the catalytic reaction was the formation of the guanidinium–malonate ion-pair complex. The conjugate addition may proceed via conventional Brønsted acid (side-on, SO) or Brønsted–Lewis acid bifunctional activation (face-on, FO)

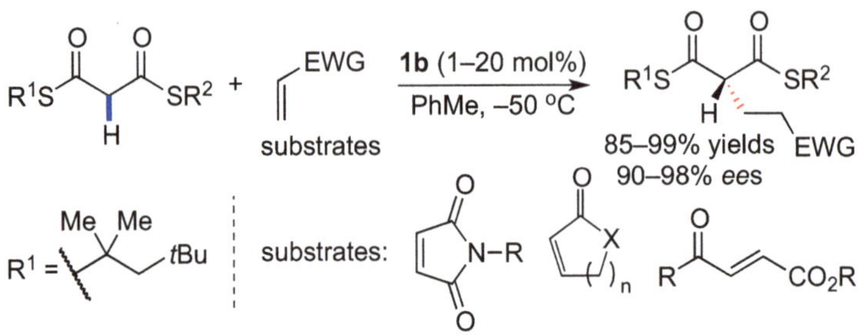


**Scheme 3** Asymmetric Michael reaction of 1,3-thiomalonates

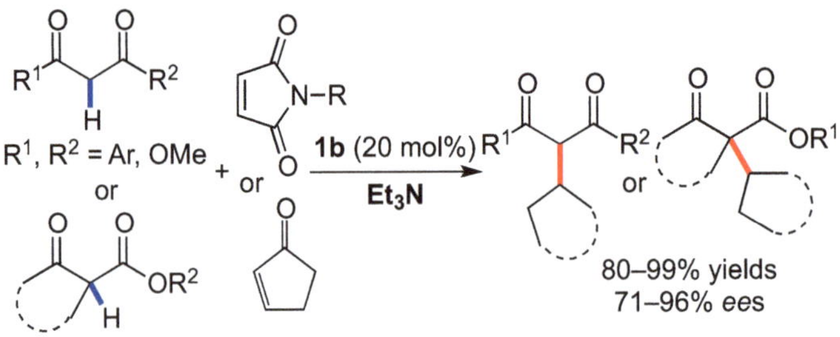


**Scheme 4** Enantioselective Michael reaction of broadly useful 1,3-dicarbonyl donors

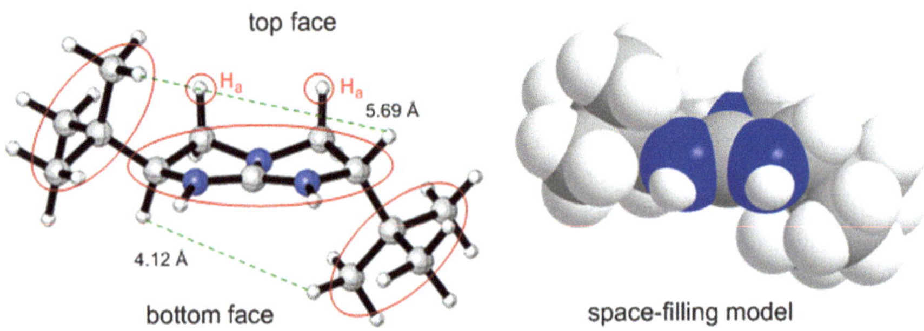


**Fig. 3** Optimised geometry of protonated guanidine **1b** (reprinted with permission from [29])

(Fig. 4). The *R*-TS-SO1 was calculated to be the lowest energy amongst all eight transition states. Hence, the conventional bifunctional activation mode is energetically more favourable. The calculated enantiomeric excess using activation free energies is in good agreement with experimental result.

The Diels–Alder reaction is typically catalysed using a Lewis acid or Brønsted acid (for reviews on asymmetric Diels–Alder reactions, see [30, 31]). Brønsted base-catalysed Diels–Alder reactions are relatively less explored. Chiral bicyclic guanidine **1c** was found to catalyse the formal Diels–Alder reaction between anthrones and maleimides with excellent enantioselectivities (Scheme 5) [32].

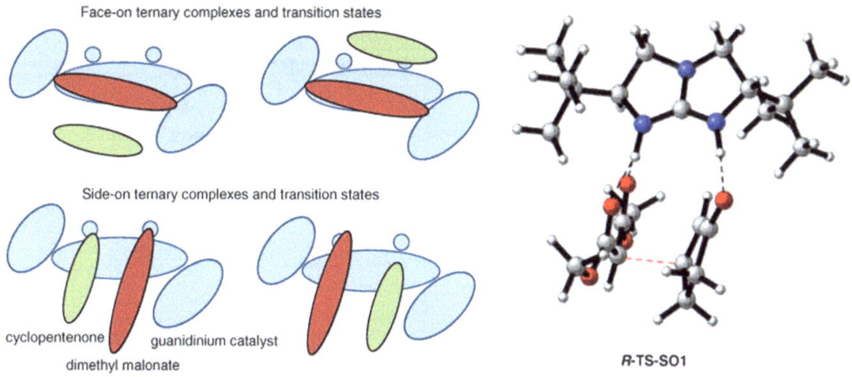


**Fig. 4** Lowest energy transition state (reprinted with permission from [29])

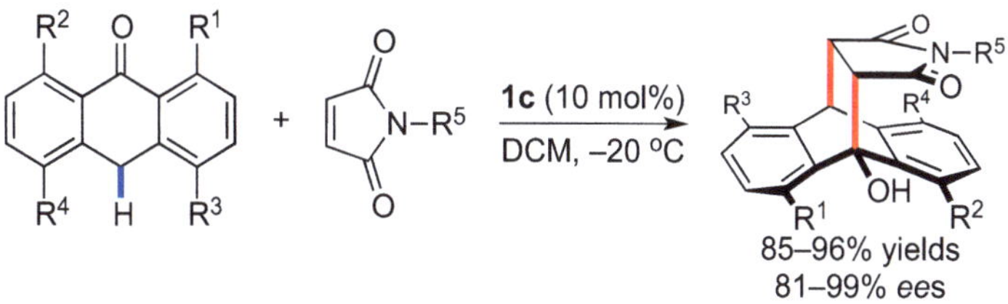


**Scheme 5** Diels–Alder reactions between substituted anthrones and maleimides

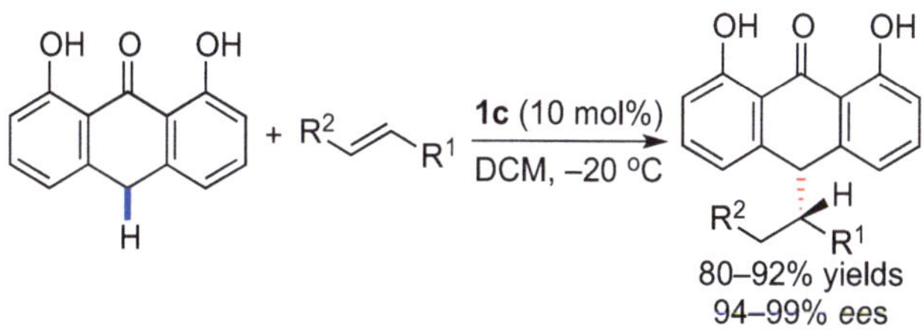


**Scheme 6** Enantioselective addition reactions of dithranol

Prolonged stirring or base treatment of the reaction product would induce a *retro*-aldol ring opening reaction. However, when dithranol (1,8-hydroxyanthrone) was used as the dienophile, with various electron-withdrawing alkenes, in the presence of catalyst **1c**, the reactions always led to the exclusive formation of the Michael addition product (Scheme 6). We also prepared chiral *N*-hydroxyphthalimide (NHPI) derivatives by adding anthrones to *N*-acetoxymaleimide [33] and utilised these novel chiral NHPI as catalyst for several asymmetric aerobic oxidation reactions.

We examined the catalytic activities of TBD **9** in phospha-Michael reactions using diphenylphosphine oxide and phosphite nucleophiles [34]. Subsequently, we investigated the asymmetric versions using chiral bicyclic guanidine **1b** [35]. It was

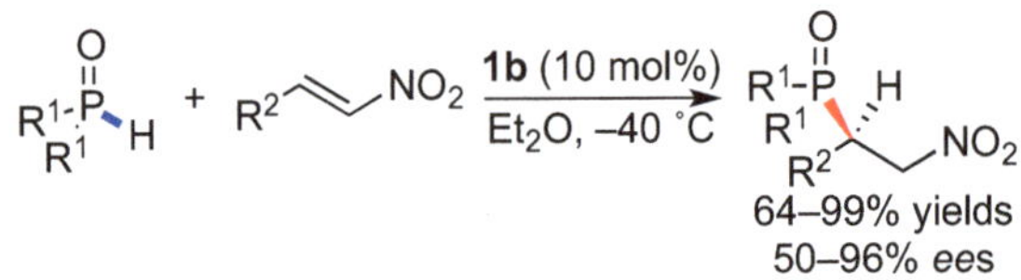


**Scheme 7** Phospha-Michael reactions of β-nitrostyrenes

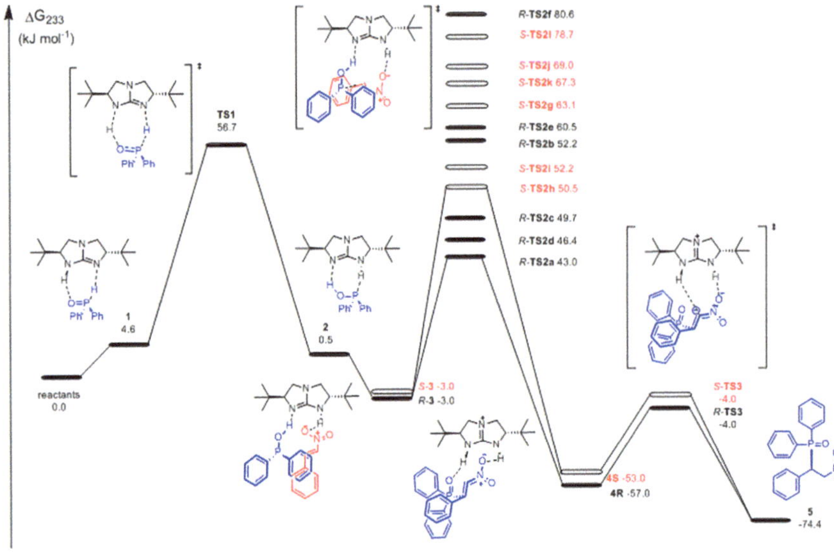


**Scheme 8** Schematic energy profile of phospha-Michael reaction (reprinted with permission from [36])

found that guanidine **1b** worked well in the presence of reactive β-nitrostyrene derivatives and secondary phosphine oxides (Scheme 7). Other phosphine-containing donors such as phosphite derivatives did not work particularly well, presumably due to their higher $pK_a$. One of the disadvantages of this reaction was the limited substrate scope of the secondary phosphine oxide derivatives that can be used to achieve high levels of enantioselectivities. However, many adducts that were obtained with moderate enantioselectivities can be isolated as pure products in excess of 99% *ee* through a simple recrystallization.

We investigated the reaction mechanism using computational tools and found that guanidine catalyst **1b** acts as both Brønsted base and acid (Scheme 8) [36]. In the first step, the catalyst tautomerises the phosphine oxide to phosphinic acid. Next, the phosphinic acid tautomer serves as the nucleophile. The bifunctional guanidine catalyst activates both reactants simultaneously via hydrogen bonding in the transition state. The calculated activation barriers favours the *R*-enantiomer of the product, and this result correlates well with the experimental observation. In the last step, the N–H proton from the catalyst is transferred to the α-carbon of the anionic nitroalkane.

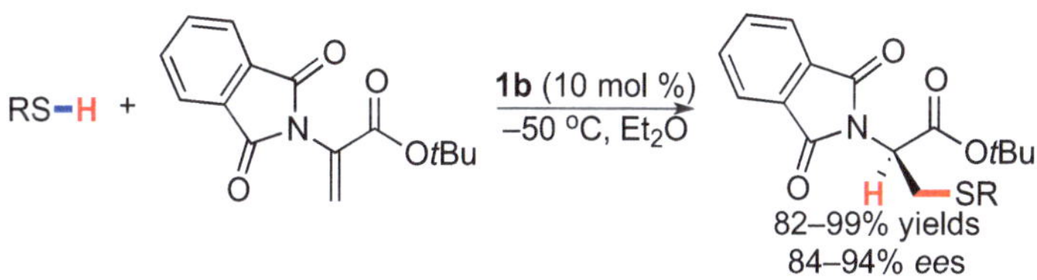


**Scheme 9** Tandem conjugate addition–enantioselective protonation of 2-phthalimidoacrylate

## 2.3 Brønsted–Lewis Acid Bifunctional Activation

After we had success with the initial projects, we began to wonder what type of reactions would be the niche reactions of Brønsted base catalyst. We found that we can shuttle one of the smallest functional groups, a proton, enantioselectively with a Brønsted base catalyst (for selected reviews on enantioselective protonation, see [23, 37]). The reaction that allows us to realise this point is the tandem conjugate addition–enantioselective protonation of 2-phthalimidoacrylate (Scheme 9) (for a review on synthetic applications of 2-phthalimidoacrylates, see [38, 39]). The reaction provided enantio-enriched cysteine analogues, and to attain high enantioselectivities, the *tert*-butyl ester group was essential [40]. In collaboration with Wong, it was found computationally that the conjugate addition of a thiophenolate anion to *s-cis*-2-phthalimidoacrylate would lead to the *E*-enolate, while the *s-trans* conformation would give the *Z*-enolate. We also found that the *s-cis* conformation was energetically more favourable as it faced less steric repulsion and formed stronger hydrogen bonds in the transition state. Surprisingly, we found an unconventional mode of non-covalent interaction (Scheme 10, pathway B); Brønsted–Lewis acid bifunctional activation was found to be more energetically feasible. The alkene interacted with the guanidinium catalyst via dual hydrogen bonds, while the thiophenolate was bound to the electrophilic central carbon of guanidinium cation via Lewis acid interaction. Due to the strength of the guanidinium ion-pair interaction, only the *Si* face was attacked in both the C–S bond forming and enantioselective protonation steps.

We went on to explore the enantioselective protonation of cyclic itaconimides, which form the *Z*-enolate intermediate exclusively (Scheme 11) [39, 41, 42]. We found that *N*-mesityl itaconimide was needed to achieve high levels of enantioselectivity. Other than thiols, secondary phosphine oxides were excellent nucleophiles for this reaction, attaining excellent enantioselectivities. A large substituent such as a *tert*-butyl group on the *ortho*-position of the *N*-phenyl ring of the itaconimides restricted rotation about the C–N axis and lead to a mixture of diastereoisomeric atropisomers (Scheme 12) [41, 42]. The *anti* diastereoisomer was shown to have a higher *ee* than the *syn* isomer.

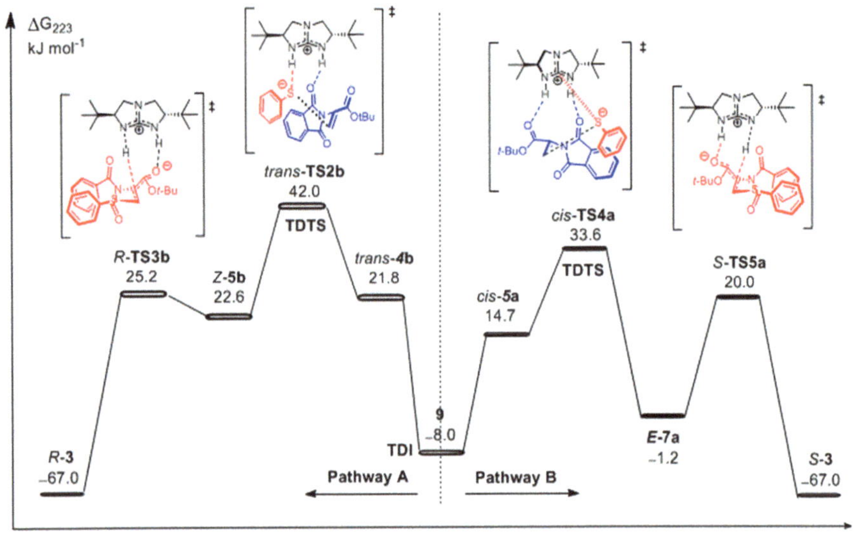


**Scheme 10** DFT studies of the enantioselective protonation of 2-phthalimidoacrylate (reprinted with permission from [40])

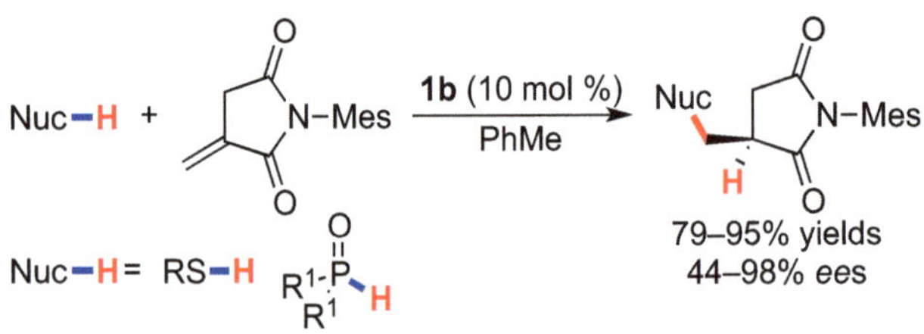


**Scheme 11** Enantioselective intermolecular proton shuttling of itaconimides

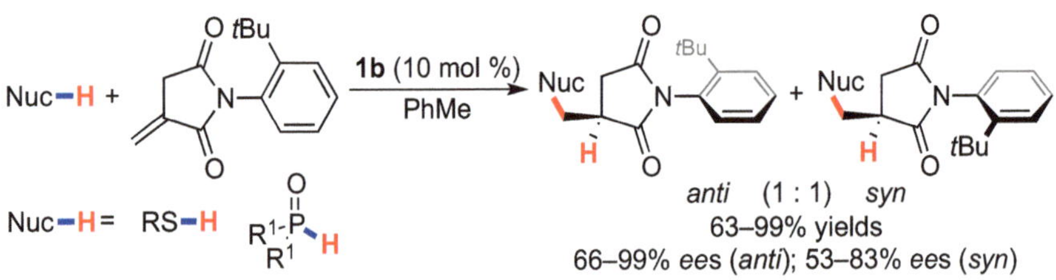


**Scheme 12** Enantioselective protonation of axially chiral itaconimide

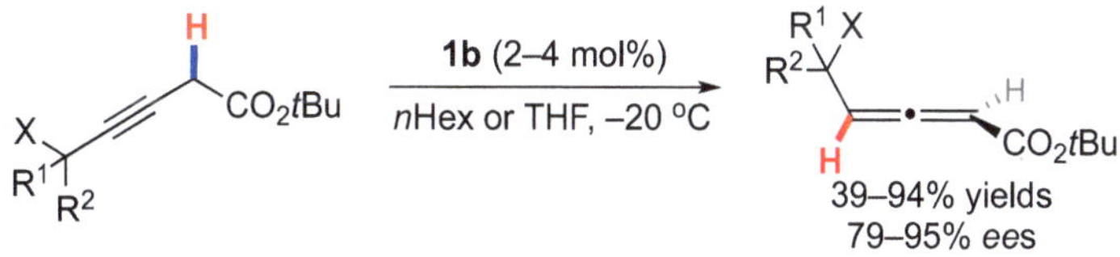


**Scheme 13** Enantioselective intramolecular proton shuttling

## *2.4 Isomerisation Reactions*

The isomerisation of alkynes is one of the most atom-economical methods to prepare allenes (Scheme 13) [43]. Here, the Brønsted base catalyst abstracts a proton from the α-carbon of 3-alkynoate derivative to generate a prochiral enolate–alkyne intermediate. As the enolate–alkyne rearranges, it takes back a proton from the catalyst to form a chiral allene. Overall, a proton is shuttled in an intramolecular 1,3-proton shift fashion, mediated by the Brønsted base catalyst. The substrate scope was wide and included both aryl- and alkyl-substituted 3-alkynoate derivatives. We studied the energy profile of this reaction using DFT calculations (B3LYP) and found that the allenes were more stable than the corresponding alkynes by more than 3.0 kcal $mol^{-1}$. Hence, the reaction was thermodynamically favourable, but had a high activation barrier. Since most of the chiral allenoates were inseparable from the 3-alkynoate derivatives by column chromatography, the axial chirality of the allenoates was directly transformed into products with point chirality with minimal loss of enantioselectivities. Subsequently, we also developed a tandem alkyne isomerisation–Michael reaction catalysed by guanidine **9** for the synthesis of oxa and aza cycles [44].

Next we investigated the direct Brønsted base-catalysed enantioselective deuteration of enolates formed under equilibrium conditions. In the presence of a chiral Brønsted base catalyst, ketones underwent rapid base-catalysed enolization. Deuteration of the enolate by the chiral catalyst gave an enantio-enriched deuterated ketone. Due to kinetic isotope effects, the reverse reaction would be less favourable. Using α-fluorinated ketones, which had a compatible $pK_a$ with catalyst **1b**, the H–D exchange reaction gave up to 30% *ee* (Scheme 14) [45]. DFT calculations suggested a water-assisted mechanism in the deuteration step (Fig. 5). We also determined that as the kinetic isotope effect increased, enantioselectivities would also increase. We also observed that when the starting material was fully deuterated, the enantioselectivity value started to erode and a racemate was formed.

During our investigation of the enantioselective protonation of itaconimides, we observed that isomerisation of the alkene to the more stable maleimide derivatives can be catalysed by Brønsted base. This led us to study imino ene-type reaction of alkylidene-succinimide derivatives (Scheme 15) [46]. One of the crucial factors in ensuring high enantioselectivities was the use of *N*-3-ethylpentan-3-yloxycarbonyl (Eoc)-protected imines. This protecting group was slightly bulkier than Boc. The

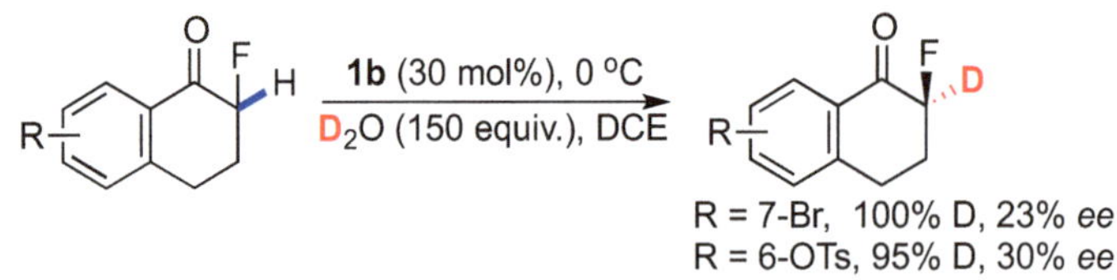


**Scheme 14** Enantioselective deuterium exchange of α-fluorinated ketones

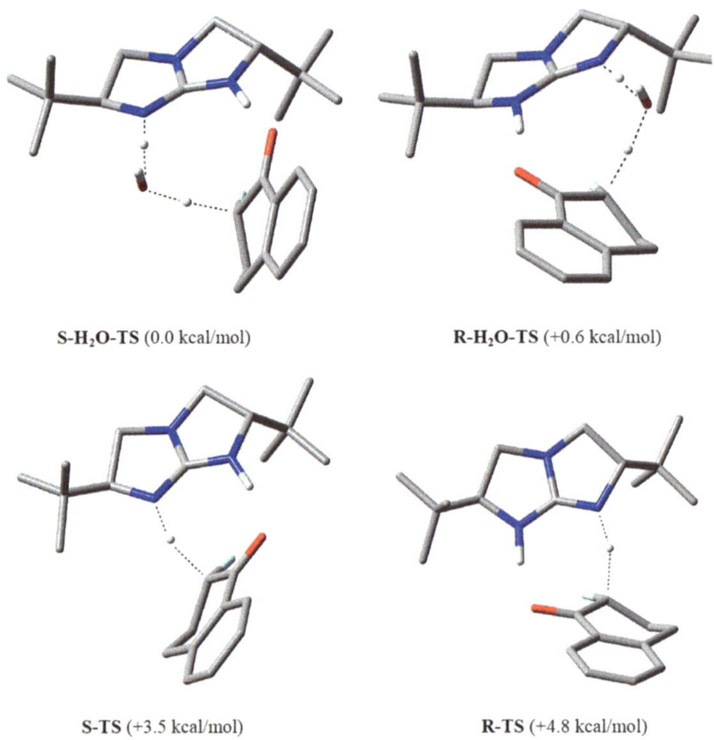


**Fig. 5** Water-assisted deuteration of cyclic ketone enolates (reprinted with permission from [45])

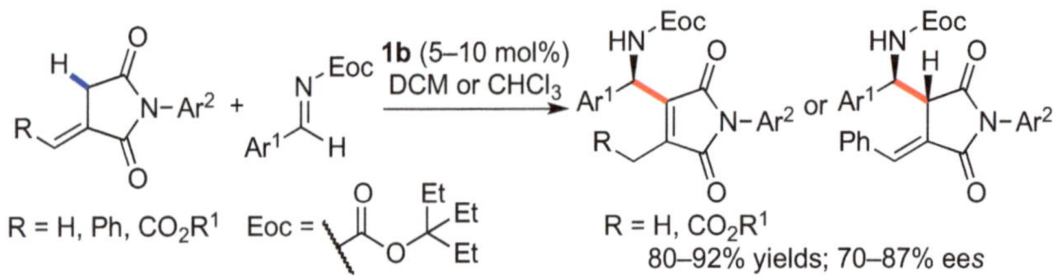


**Scheme 15** Enantioselective imino ene reaction of alkylidene-succinimide derivatives

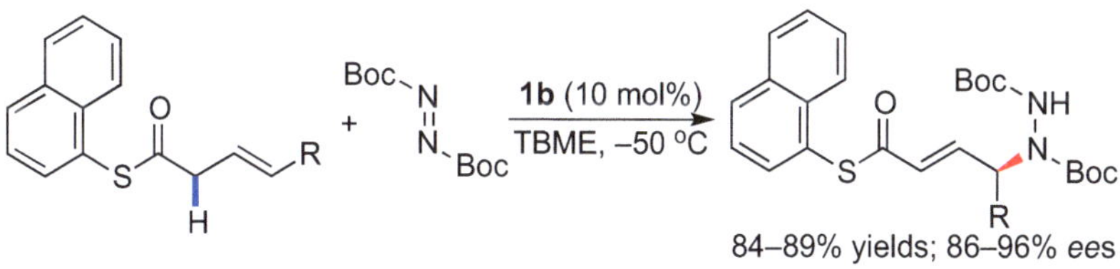


**Scheme 16** Enantiodivergent γ-amination of (*E*)-β,γ-unsaturated thioesters

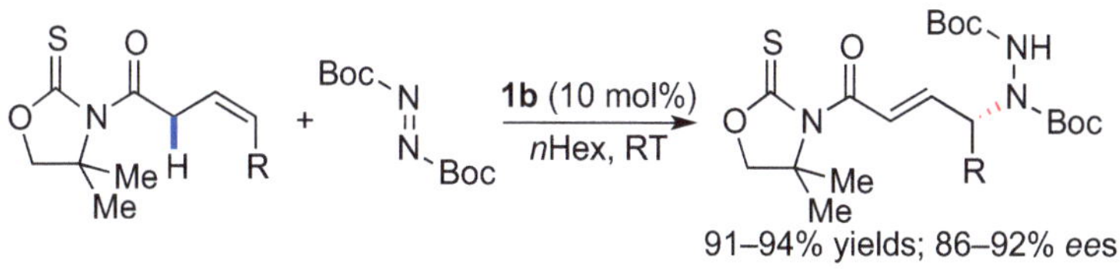


**Scheme 17** Enantiodivergent γ-amination of (*Z*)-β,γ-unsaturated thioesters

regioselectivity was excellent since no γ-addition product was observed in all examples examined. The imino ene-type α-addition products were obtained in moderate to good enantioselectivities. Deuterium-labelling experiments indicated a 1,3-proton shift during the reaction.

In a follow-up work, we explored the γ-functionalisation of acyclic substrates such as β,γ-unsaturated carbonyl compounds. We hypothesised that a vinyl group might decrease the $pK_a$ of the protons on the α-carbon and thus facilitate deprotonation. We further increased the acidity of the protons through the use of thioesters. The reaction between (*E*)-β,γ-unsaturated thioesters and di-*tert*-butyl azodicarboxylate went smoothly to give the γ-amination products (Scheme 16) [47]. The interesting part of this work was that this reaction was stereospecific, or rather enantiodivergent, as both enantiomers could be obtained from same chiral catalyst. When we switched to (*Z*)-β,γ-unsaturated thioesters, the γ-amination products had the opposite configuration (Scheme 17). The 4,4-dimethyloxazolidine-2-thione group was the optimal auxiliary in this case. Adducts with three consecutive chiral centres, consisting of 1,2-diol-3-amino groups, were obtained using a diastereoselective $OsO_4$-catalysed dihydroxylation of the γ-amination products.

We studied the origin of enantioselectivity using DFT calculations (Fig. 6) [47]. Side-on transition states are favoured due to strong intermolecular interactions between catalyst and azodicarboxylate. The *s-trans* dienolates had lower energies than *s-cis* dienolates as they did not experience any 1,3-allylic interaction. The more stable transition states indicated that the 1-naphthyl substituent was pointing away from the *tert*-butyl substituent of the catalyst. As the newly formed chiral centre was at the alkene carbon bearing the substituent, a different *E*/*Z* alkene configuration would result in the opposite configuration of the new chiral centre.

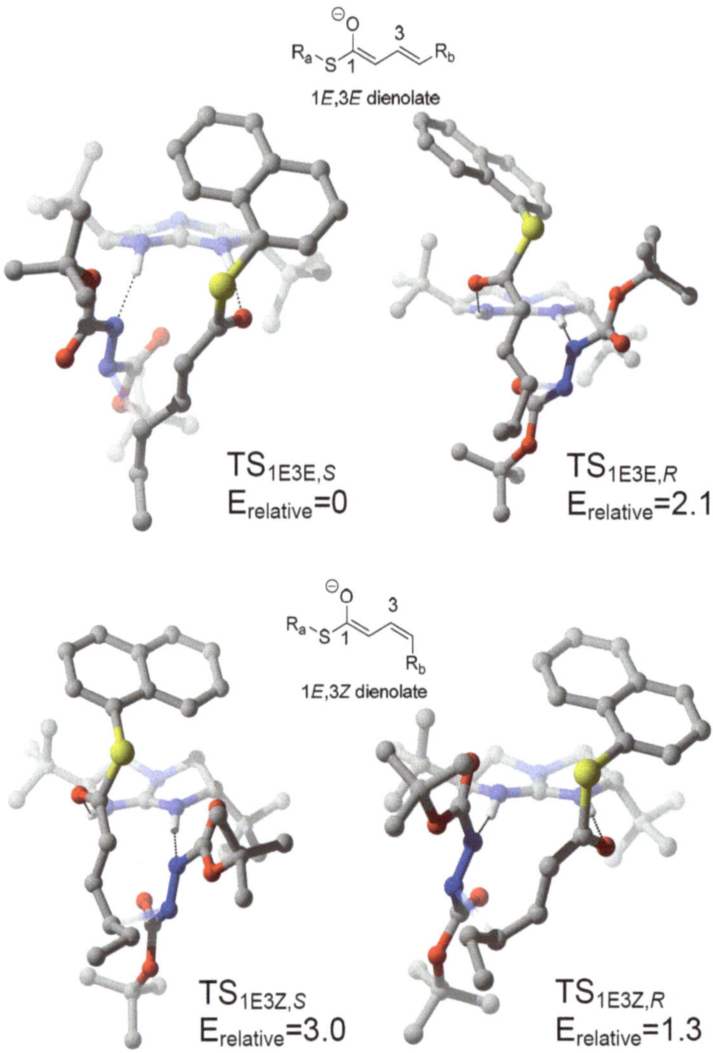


**Fig. 6** Most energetically favoured transition states derived from *s-trans* dienolate (reprinted with permission from [47])

## 2.5 Late Developments

In the later stages of the development of conjugate reactions, we looked at other activated alkenes. We went on to expand the substrate scope to include other acyclic Michael acceptors such as (*E*)-4-oxo-4-arylbutenamides (Scheme 18) [48]. We screened various auxiliary groups and found 2-oxazolidinone worked well in this case.

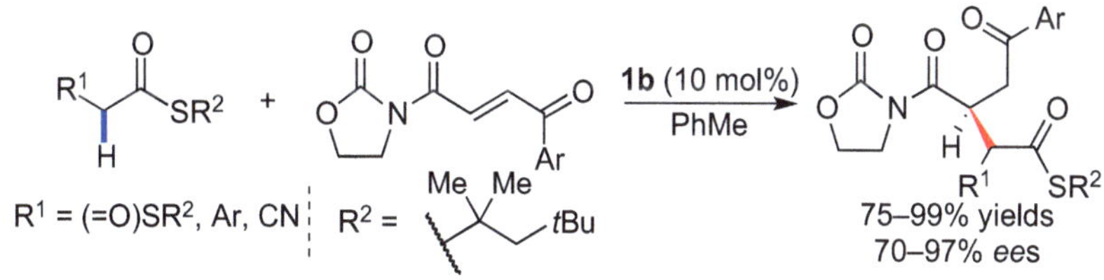


**Scheme 18** Asymmetric Michael reaction of (*E*)-4-oxo-4-arylbutenamides

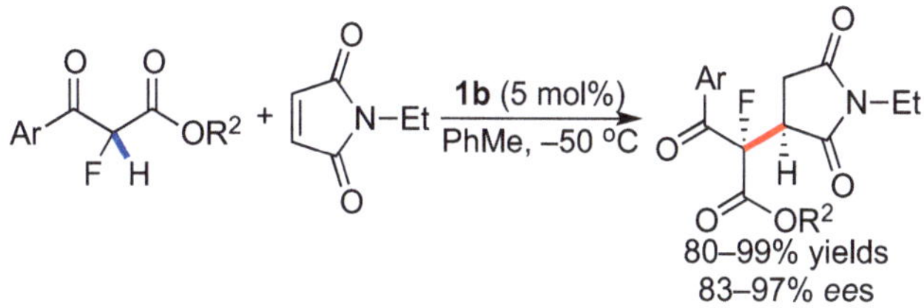


**Scheme 19** Asymmetric synthesis of fluorinated quaternary carbon centre

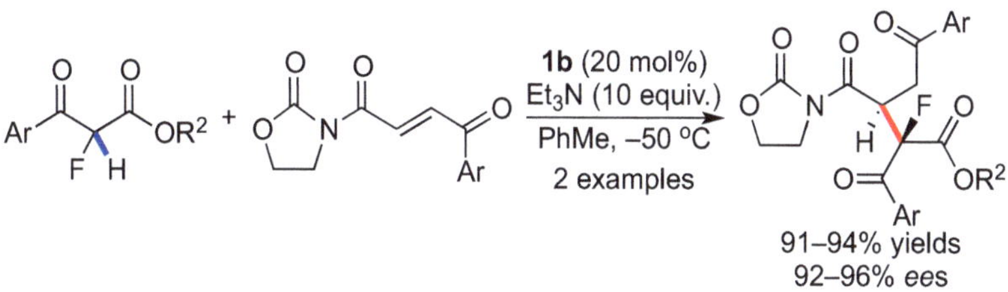


**Scheme 20** Enantioselective Michael reaction of fluoro-containing nucleophiles with acyclic Michael acceptors

We also became interested in the preparation of chiral fluorinated compounds, as they are extremely useful in medicinal chemistry. However, compounds containing fluorinated quaternary carbon centre are uncommon. One simple approach towards these molecules is to introduce fluorine into the target molecule by using α-fluoro-β-ketoesters as the nucleophile. In the presence of the fluoro group, the increased acidity of the α-proton of the β-ketoester renders it a nucleophile with higher reactivity. Both the obtained enantioselectivities and diastereoselectivities were very high (Scheme 19) [49]. The reduction of the ketone moiety in the final product gave an enantiopure fluorohydrin as a single diastereomer with three contiguous chiral centres. Although α-fluoro-β-ketoesters add more slowly to acyclic Michael acceptors, the use of triethylamine as an additive accelerated reaction rates (Scheme 20).

DFT calculations were again used as the tool to gain insights into the reaction mechanism [49]. The α-fluoro-β-ketoester existed as a stable enol; the guanidinium

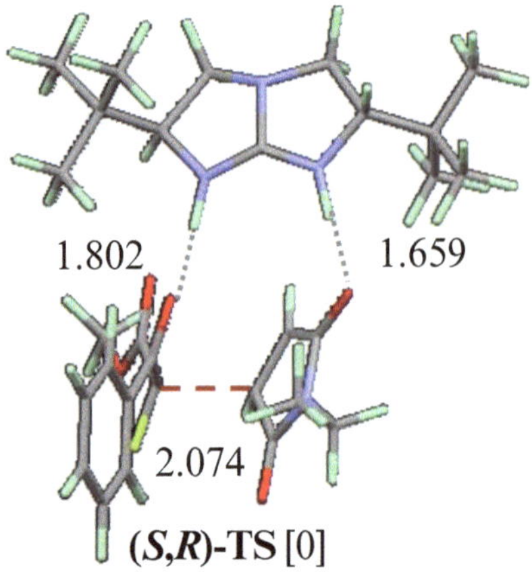


**Fig. 7** Optimised geometry of the most stable TS leading to (*S*,*R*) stereoisomer (reprinted with permission from [49])

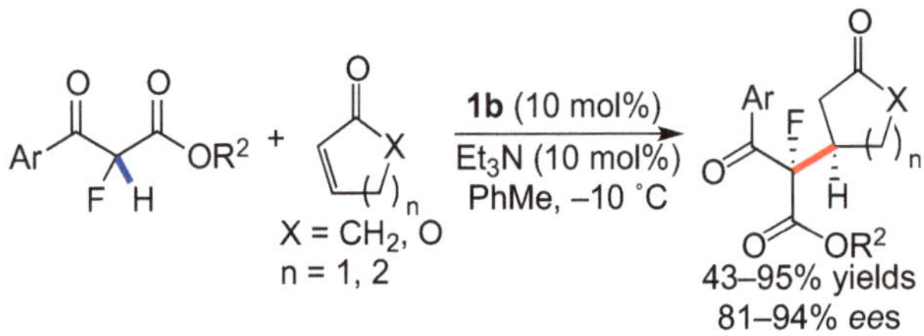


**Scheme 21** Enantioselective Michael reaction of fluoro-containing nucleophiles with cyclic enones

cation and α-fluoro-β-ketoester anion complex were held by dual hydrogen bonds. Next, the complex formed a pre-transition state (pre-TS) complex with maleimide in either a face-on or a side-on transition state. Since guanidine **1b** was bifunctional, the side-on transition state was preferred due to strong hydrogen bonding with the carbonyl group of maleimide. The most stable TS led to the formation of the (*S*,*R*) stereoisomer which correlated well with the experimentally observed results (Fig. 7). Subsequently, in collaboration with Jiang's group, we extended the same concept to cyclic enones as Michael acceptors (Scheme 21) [50].

Next, we investigated the Mannich reaction between α-fluoro-β-ketoesters and various protected imines (Scheme 22) [51]. Both the *N*-3-ethylpentan-3-yloxycarbonyl (Eoc) imine and the 4,4-dimethyl-oxazolidin-2-one auxiliary were critical for excellent enantioselectivities. In addition, both α-fluoro-β-ketosulfones and α-fluoro-β-nitrosulfones were compatible with the catalytic Mannich reaction. The oxazolidinone moiety in the product could easily be converted into an ester group using potassium carbonate in alcoholic solvents. Base hydrolysis and decarboxylation lead to α-fluoro-β-amino ketones. *Retro*-Claisen condensation of the product gave *syn*-α-fluoro-β-amino esters as the major diastereoisomer.

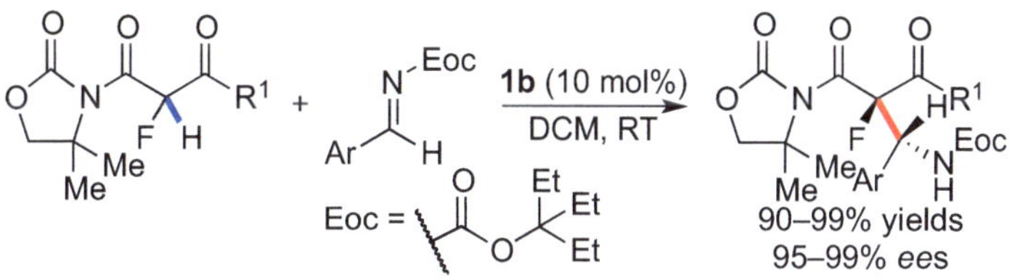


**Scheme 22** Asymmetric Mannich reaction of fluoro-containing nucleophiles

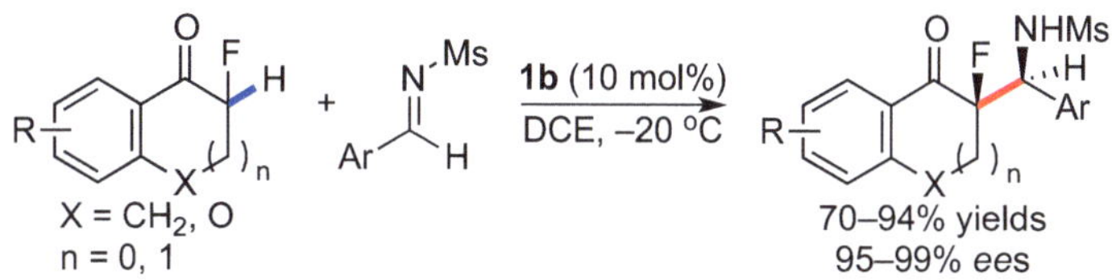


**Scheme 23** Synthesis of chiral α-fluorinated-β-amino aromatic cyclic ketones

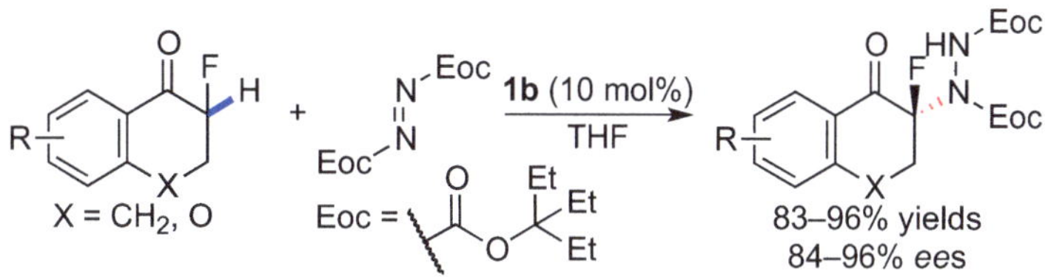


**Scheme 24** Enantioselective direct α-amination of fluoro-containing nucleophiles

We found that cyclic fluorine-containing nucleophiles also work well for Mannich reactions (Scheme 23) [52]. The *syn* diastereomer was formed as a major product with excellent enantioselectivity. While poor enantioselectivity was observed for the *anti*-isomer, the two diastereoisomers were easily separated by flash chromatography. Direct α-amination of α-fluorinated aromatic cyclic ketones allowed us to access unique fluorinated α-amino cyclic ketones. The protecting group of the azodicarboxylate should be huge, such as in di-3-ethylpentan-3-yl azodicarboxylate, in order to meet the steric requirements. In general, excellent enantioselectivities and yields were obtained for the α-aminated products (Scheme 24). This work leads us to review the use of organofluoro nucleophiles in asymmetric catalysis (for a review on asymmetric synthetic applications of organofluoro nucleophiles, see [53]).

Decarboxylative reactions of malonic acid half thio/oxyesters are biomimetic reactions. Similar reactions are used by nature for polyketide synthesis (for a review on catalytic decarboxylative reactions, see [54]). We reported the asymmetric decarboxylative Mannich reaction using sterically hindered malonic acid half thioesters (Scheme 25) [55]. Using a similar strategy, we also found that the decarboxylative α-amination reaction worked well (Scheme 26). This reaction is

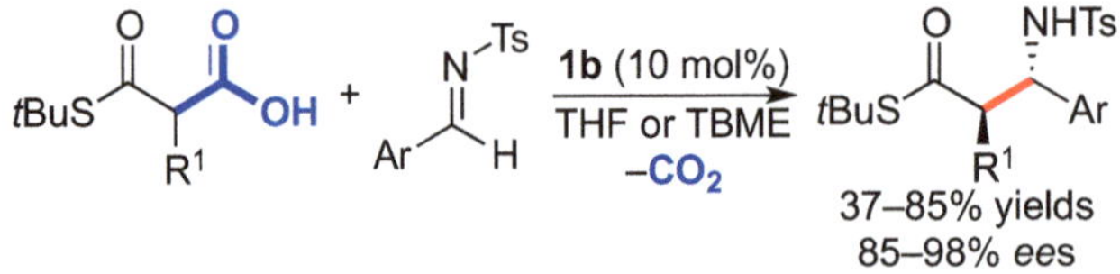


**Scheme 25** Enantioselective decarboxylative Mannich reactions

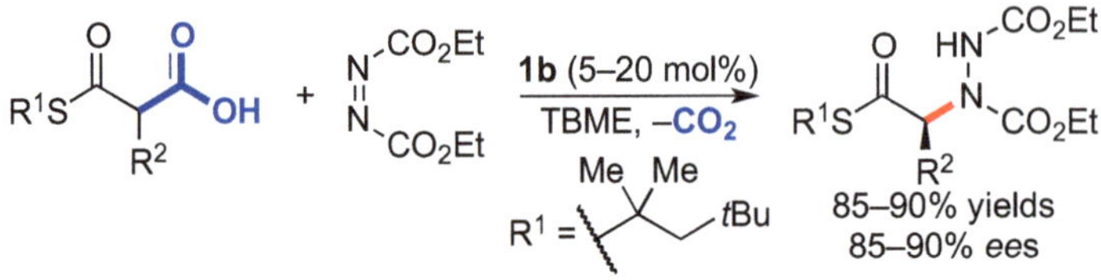


**Scheme 26** Enantioselective decarboxylative α-amination reactions

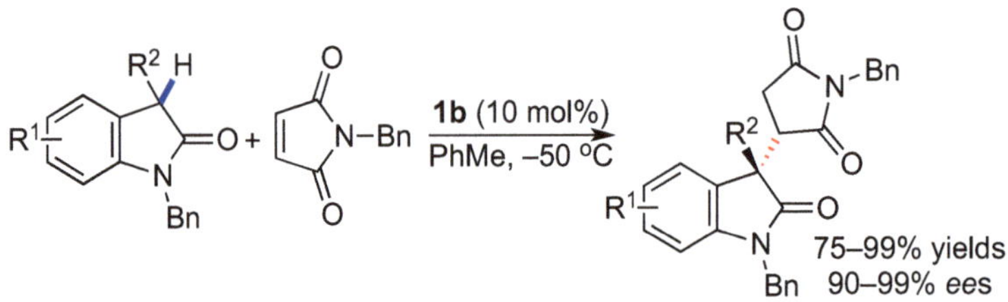


**Scheme 27** Syntheses of chiral 3-substituted oxindoles via conjugate addition

useful for preparing α-amino acid derivatives. Our computational results strongly indicated that Mannich addition occurs before decarboxylation, and the rate-determining step is the decarboxylation step. We detected the Mannich addition anionic intermediate, the intermediate before decarboxylation, using electrospray ionisation (ESI) mass-spectrometric analysis.

In collaboration with Jiang, we found that 3-substituted oxindoles, a privileged heterocyclic motif, add to *N*-maleimides (Scheme 27) [56]. The reaction worked for quite a number of *N*-maleimides and a range of different substituted benzyl groups. As an extension, we also investigated the conjugate addition of 3-benzyl-substituted oxindoles to cyclic enones (Scheme 28) [57].

However, when we used (*E*)-4-oxo-4-arylbutenones as the acceptor for conjugate addition reaction, we encountered problems. [48]. Under Brønsted base catalysis, significant decomposition of the dithiomalonate occurred to give free thiols and a sulfa-Michael reaction was observed. Thus, we reduced the basicity of the reaction condition to suppress the side product formation (Scheme 29) [48]. Under phase-transfer conditions with a mild base, both symmetrical and unsymmetrical (*E*)-1,4-unsaturated diketones gave high regioselectivities and enantioselectivities.

We subsequently extended this approach to the asymmetric phase-transfer alkylation of 3-substituted-2-oxindole derivatives (Scheme 30) [58]. Using the

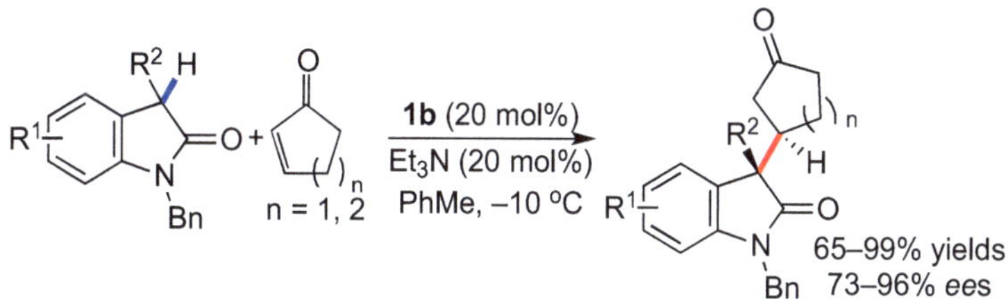


**Scheme 28** Asymmetric conjugate additions of 3-substituted oxindoles to cyclic enones

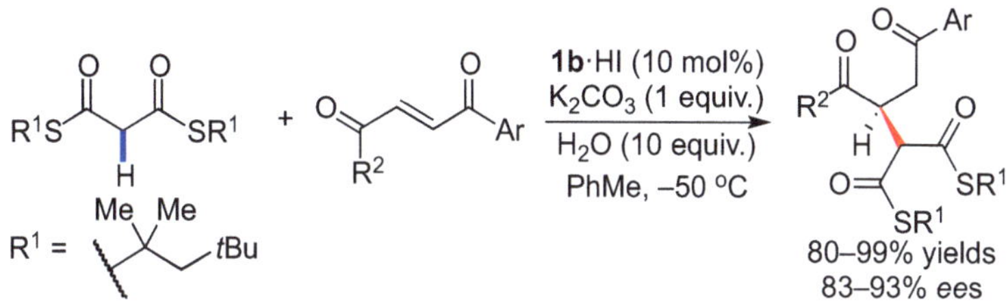


**Scheme 29** Asymmetric Michael reactions of 1,4-unsaturated diketones

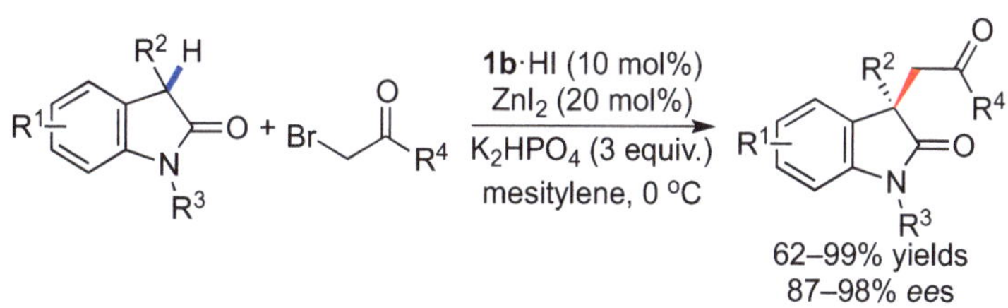


**Scheme 30** Enantioselective phase-transfer alkylation of 3-substituted-2-oxindoles

bicyclic guanidinium iodide salt of **1b** and the cooperative effect of $Zn^{2+}$ as a Lewis acid were both vital for this reaction to work. The substrate scope is broad and includes various substituted 2-oxindoles as well as ester-, amide- and ketone-activated bromomethane alkylating reagents. Finally, we applied this methodology to the synthesis of bioactive pyrrole– and furan–indolines.

# 3 Other Chiral Guanidines as Brønsted Base Catalysts

The field of chiral guanidine catalysis is rapidly evolving, and we now have more varieties of guanidine catalysts with novel designs (Fig. 8) (for selected recent references on novel chiral guanidine catalysts, see [59–65]). In order to advance the

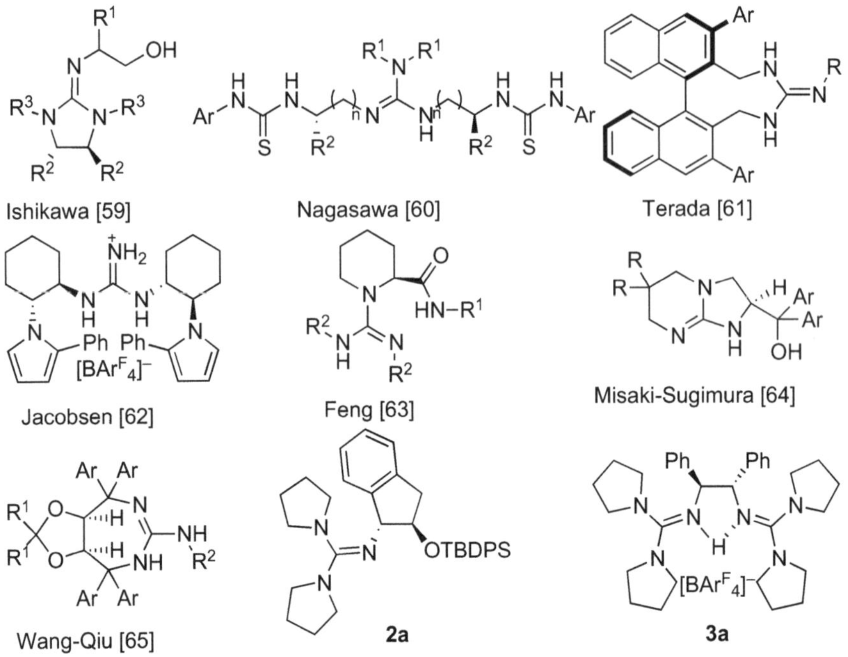


**Fig. 8** Selected examples of chiral guanidine and guanidinium catalysts

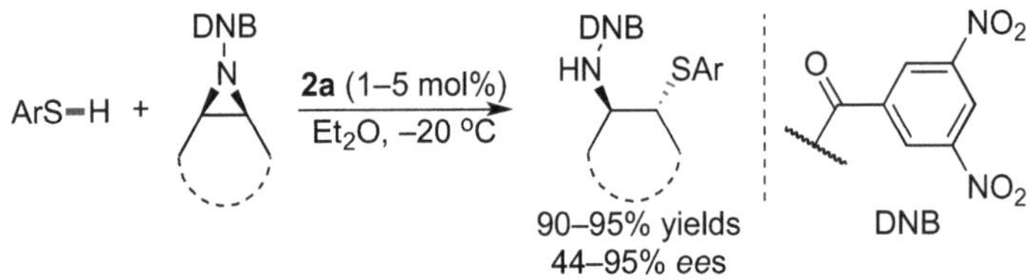


**Scheme 31** Enantioselective desymmetrization of *meso*-aziridines

field, we still need to find guanidine catalysts that are easy to prepare and exhibit high reactivity as well as wide reaction compatibility.

We reported the enantioselective desymmetrization of *meso*-aziridines catalysed by guanidine **2a** (Scheme 31) [66]. This catalyst was easily derived from chiral aminoindanol in two simple steps. Amongst the aminoindanol-based catalysts tested, we found that TBDPS-protected **2a** gave us the best enantioselectivity and yield. Next, we examined various protecting groups on the aziridine and found that 3,5-dinitrobenzoyl was optimal. However, the reaction was quite sensitive with respect to the thiols used, and only the less reactive and bulky 2,6-dichlorothiophenol gave excellent enantioselectivities.

We also attempted to use carbamodithioic acid as a nucleophile but the results were far from ideal. However, when we generated carbamodithioic acid in situ via a

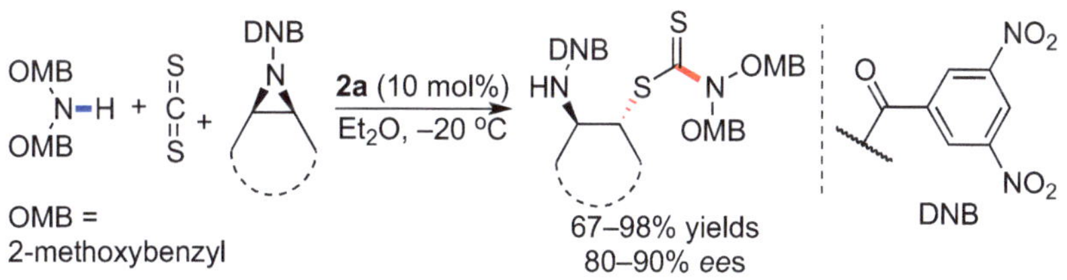


**Scheme 32** Enantioselective desymmetrization of *meso*-aziridines with in situ-generated carbamodithioic acid

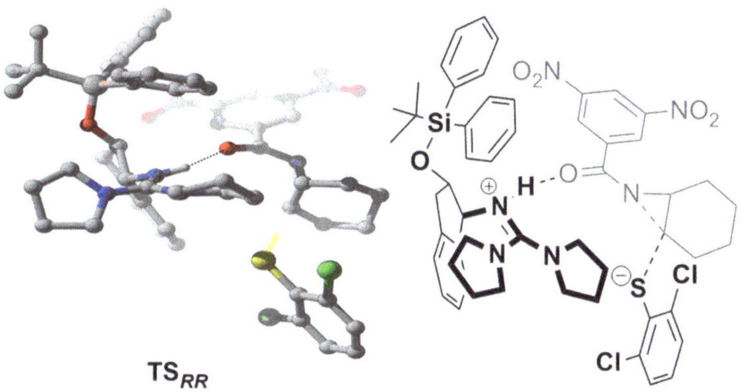


**Fig. 9** Lowest energy transition state structure for (*R*,*R*) product (reprinted with permission from [66])

three-component reaction with carbon disulphide, bis(2-methoxybenzyl)amine and aziridine, excellent enantioselectivities were obtained (Scheme 32). We used the chiral adduct obtained to prepare a zwitterionic chiral β-amino sulfonic acid. Initial DFT studies indicated that guanidine **2a** activates the carbonyl group of the aziridine instead of the aryl thiol (Fig. 9). When the thiol was substituted with chlorine at the 2- and 6-position, there was an increase in steric repulsion between the two chlorine atoms and the catalyst–aziridine complex.

## 4 Chiral Bis-Guanidinium Salts as Brønsted Acid–Brønsted Base Dual Functional Catalysts

In order to investigate new modes of reactivity, we became interested in preparing and using bis-guanidines as catalyst. We initially hypothesised that two guanidine functional groups might have a cooperative effect and hence increase the basicity of the Brønsted base. However, we found that this was not the case. Instead, we found that the reactivity of the catalyst can be tuned by varying the number of equivalents

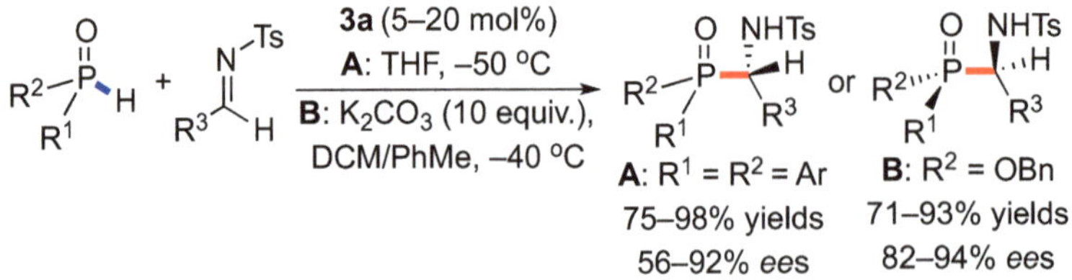


**Scheme 33** Enantioselective phospha-Mannich reactions of secondary phosphine oxide H-phosphinate derivatives

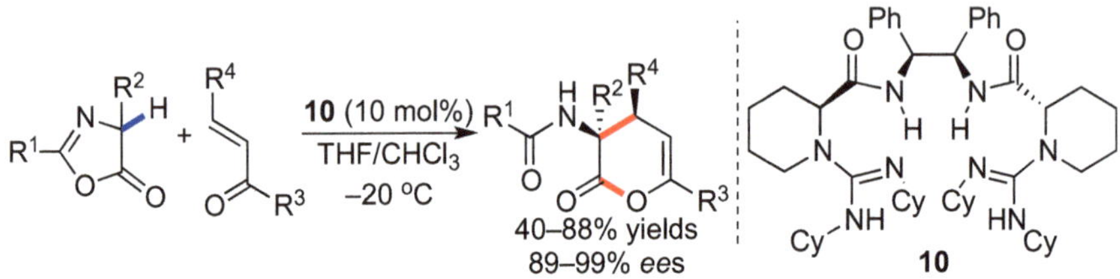


**Scheme 34** Chiral inverse electron demand hetero Diels–Alder reaction

of the acid added and the type of counter anion used. Bis-guanidinium salt **3a** (Fig. 8) was prepared by mixing the bis-guanidine and bis-guanidinium.2H$^+$ salt in a 1:1 ratio.

We wanted to investigate phospha-Mannich reactions, but catalyst **1b** was found to be unsuitable. On the other hand, however, we found that the BARF salt of bis-guanidinium **3a** worked very well for phospha-Mannich reactions (Scheme 33) [67]. Various imines reacted well with 1-naphthyl-substituted phosphine oxide, but other secondary phosphine oxide derivatives gave only moderate enantioselectivities. This reaction was also applicable to racemic H-phosphinate substrates in the presence of $K_2CO_3$, which resulted in both good diastereoselectivities and enantioselectivities. We also demonstrated the kinetic resolution of a *P*-chiral centre using a racemic H-phosphinate derivative in the phospha-Mannich reaction, which occurred with a selectivity factor of 8.2.

High enantioselectivity was observed with bis-guanidinium salt **3a** (Fig. 8) containing a single proton. As the two guanidine functional groups are orthogonal and do not share the proton equally, one guanidine group thus functions as the Brønsted acid and the other Brønsted base. This dual functional mode of activation is crucible for this reaction.

Similarly, Feng and co-workers reported the inverse electron demand hetero Diels–Alder reaction of chalcones with azlactones that were catalysed by bis-guanidine **10** (Scheme 34) [68]. This bis-guanidine or its bis-guanidinium salt counterpart, as well as other similar derivatives, proved to be highly versatile in various enantioselective transformation [69–71].

## 5 Chiral Pentanidium Salts as Phase-Transfer Catalysts

Chiral pentanidium salts **4** contain 5 nitrogen atoms in conjugation (Fig. 10). The main reason why they are not more basic than normal guanidines, as we first expected them to be, is that the catalysts are orthogonal (L-shaped) due to steric reasons; i.e., the 5 nitrogen atoms are not on the same plane. The fully alkylated pentanidium salts, however, turned out to be excellent phase-transfer catalysts. While conventional phase-transfer catalysts are $sp^3$-hybridised ammonium salts, the pentanidiums are $sp^2$ hybridised. Since pentanidiums have two faces, controlling enantioselectively is expected to be harder. The catalyst synthesis was straightforward, requiring five steps with only a single flash chromatography step and a final recrystallization step. The structure of the chiral pentanidium salt was unambiguously confirmed by single-crystal *X*-ray diffraction analysis.

We first investigated pentanidiums **4** using a classical conjugate addition, the addition of a glycinate-derived benzophenone Schiff base to α,β-unsaturated compounds [59]. With pentanidium salt **4a**, the conjugate addition proceeded well with a myriad of electron-withdrawing olefins (Scheme 35) [72]. Benzophenone imines of phosphoglycine ester also worked well in the reaction with benzyl acrylate, and further transformations lead to phosphonic analogues of (*S*)-proline.

Using pentanidium salt **4b**, we achieved the chiral α-hydroxylation of oxindoles with molecular oxygen (Scheme 36) [73]. We found that under high concentration of oxygen, the reaction would produce hydroperoxide oxindoles. Under low concentration of oxygen, the hydroxyl product was obtained instead and with higher enantioselectivities. Hence, we proposed that a hydroperoxide oxindole was the key intermediate and underwent a second kinetic resolution step. We used [33] $O_2$ isotope labelling to confirm that the source of the hydroxyl group is indeed molecular oxygen. We prepared racemic hydroperoxide oxindole and performed the kinetic resolution step separately and found that the selectivity factor *S* was 5.

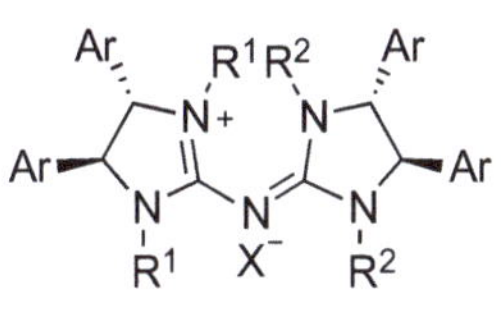


**4a**: Ar = Ph, $R^1$ = $R^2$ = Me
**4b**: Ar = Ph, $R^1$ = (4-Cl)Bn, $R^2$ = (3,5-*t*Bu)Bn
**4c**: Ar = Ph, $R^1$ = $R^2$ = (3,5-*t*Bu-2-I)Bn

**Fig. 10** Selected chiral pentanidium catalysts

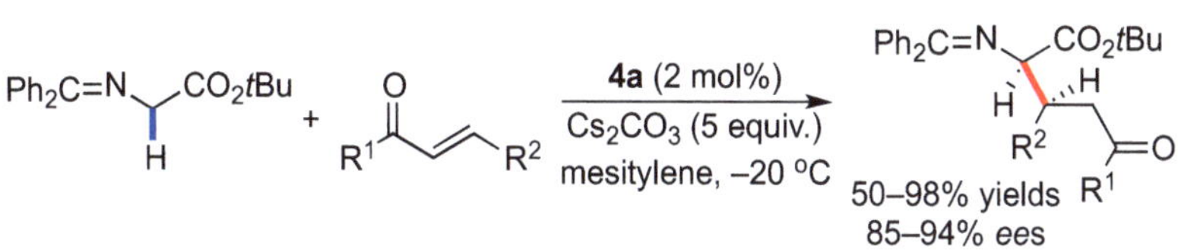


**Scheme 35** Enantioselective phase-transfer conjugate addition of glycinate-derived benzophenone Schiff base

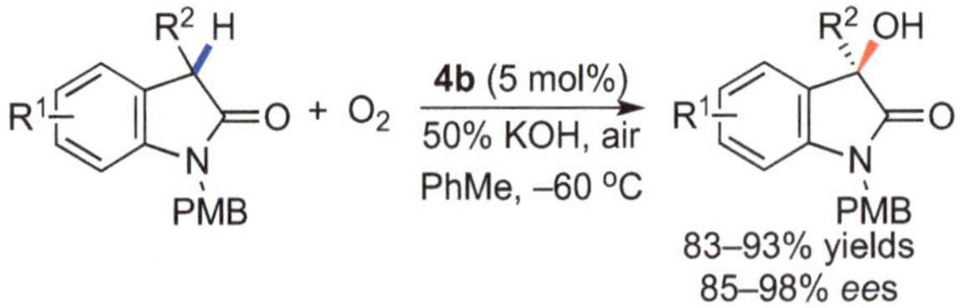


**Scheme 36** Enantioselective α-hydroxylation of oxindoles with molecular oxygen

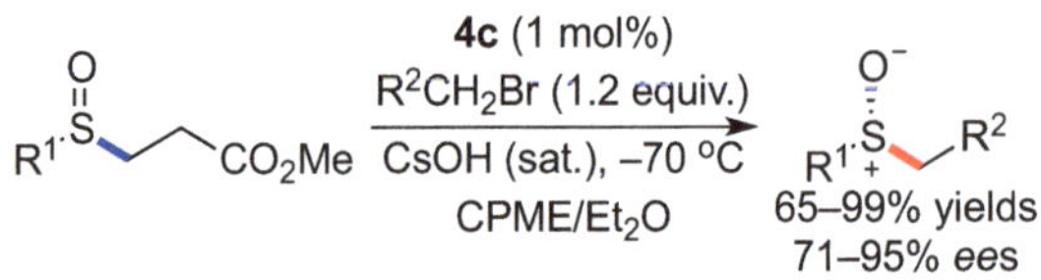


**Scheme 37** Enantioselective alkylation of sulphenate anions

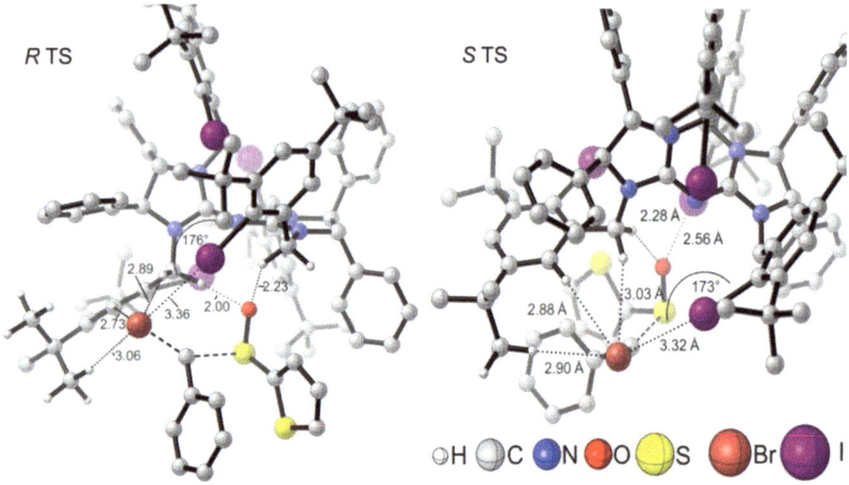


**Fig. 11** Optimised most stable transition states (reprinted with permission from [74])

Recently, we reported the asymmetric alkylation of sulphenate anions to various chiral sulfoxides using halogenated pentanidium **4c** (Scheme 37) [74]. One of the advantages of using pentanidiums is that they are highly amenable to modification, allowing for tuning of their electronic and steric properties. We introduced novel pentanidium with halogenated benzyl groups, in which the halogen groups are critical. This reaction was based on a retro-Michael reaction that was initiated under phase-transfer conditions. This generates sulphenate anion, which was subsequently alkylated by benzyl halides. A wide range of chiral heterocyclic sulfoxides was obtained using this method. Computational studies on the possible transition states of this reaction (Fig. 11) lead us to find out that (*R*)-T.S. is more stable than the corresponding (*S*)-T.S. by 1.2 kcal $mol^{-1}$ and was in good agreement with experimental observations. Another key observation was that there was evidences of Br–I halogen bonding (XB) between the leaving Br and the iodinated catalyst. There were also some nonclassical hydrogen bonds (NCHB) involved. These additional interactions helped to stabilise the transition state. This work leads us to develop bidentate dihydroimidazoline as halogen-bonding catalyst for hydrogen transfer reactions (Scheme 37) [75].

# 6 Conclusions

Brønsted base organocatalysis has made tremendous leaps forward during the past decade. Amongst the Brønsted bases, guanidine and related organocatalysts stood out prominently due to their ability to have different modes of activation. Besides dual hydrogen bonding, newer modes of interactions, such as Lewis acid activation via the electrophilic central carbon of a guanidinium cation and halogen bonding, have enriched the chemistry of guanidine and related catalysts. Currently, there is still a demand for novel and more powerful catalysts with stronger basicity. If guanidine is considered a superbase, then perhaps what we need is a 'hyperbase' that will be able to extract seemingly impossible, inert protons. Another possibility is dual catalysis of a Brønsted base catalyst with a Lewis acid or organometallic catalyst. This will open up whole new fields. We believe that this is just the beginning of an exciting future with unlimited possibilities ahead.

**Acknowledgements** We wish to thank our dedicated colleagues and collaborators whose names appeared in the citation list. This work is generously funded by National University of Singapore; Nanyang Technological University; Agency for Science, Technology and Research (A*STAR); and GlaxoSmithKline.

# References

1. Palomo C, Oiarbide M, López R (2009) Chem Soc Rev 38:632
2. Ishikawa T (ed) (2009) Superbases for organic synthesis. Wiley, Chichester
3. Schreiner PR (2003) Chem Soc Rev 32:289
4. Doyle AG, Jacobsen EN (2007) Chem Rev 107:5713
5. Takemoto Y (2010) Chem Pharm Bull 58:593
6. Corey EJ, Ohtani M (1989) Tetrahedron Lett 30:5227
7. Corey EJ, Grogan M (1999) Org Lett 1:157
8. Coles MP (2009) Chem Commun 3659
9. Leow D, Tan C-H (2009) Chem Asian J 4:488
10. Terada M (2010) J Synth Org Chem Jpn 68:1159
11. Ishikawa T (2010) Chem Pharm Bull 58:1555
12. Leow D, Tan CH (2010) Synlett 1589
13. Sohtome Y, Nagasawa K (2010) Synlett 1
14. Fu X, Tan C-H (2011) Chem Commun 47:8210
15. Selig P (2013) Synthesis 45:703

16. Echavarren E, Galan A, Lehn J-M, de Mendoza J (1989) J Am Chem Soc 109:4994
17. Galán A, Andreu D, Echavarren AM, Prados P, de Mendoza J (1992) J Am Chem Soc 114:1511
18. Gleich A, Schmidtchen FP, Mikulcik P, Müller G (1990) J Chem Soc Chem Commun 55
19. Davis AP, Dempsey KJ (1995) Tetrahedron Asymmetry 6:2829
20. Howard-Jones A, Murphy PJ, Thomas DA (1999) J Org Chem 64:1039
21. Allingham MT, Howard-Jones A, Murphy PJ, Thomas DA, Caulkett PWR (2003) Tetrahedron Lett 44:8677
22. Ye W, Leow D, Goh SLM, Tan C-T, Chian C-H, Tan C-H (2006) Tetrahedron Lett 47:1007
23. Wang C, Goh CMT, Xiao S, Ye W, Tan CH (2013) J Synth Org Chem Jpn 71:1145
24. Ye W, Xu J, Tan C-T, Tan C-H (2005) Tetrahedron Lett 46:6875
25. McKay AF, Kreling M-E (1962) Can J Chem 40:1160
26. Kiesewetter MK, Scholten MD, Kirn N, Weber RL, Hedrick JL, Waymouth RM (2009) J Org Chem 74:9490
27. Ye W, Jiang Z, Zhao Y, Goh SLM, Leow D, Soh Y-T, Tan C-H (2007) Adv Synth Catal 349:2454
28. Jiang Z, Ye W, Yang Y, Tan C-H (2008) Adv Synth Catal 350:2345
29. Wong MW, Ng AME (2014) Aust J Chem 67:1100
30. Corey EJ (2002) Angew Chem Int Ed 41:1650
31. Shen J, Tan CH (2008) Org Biomol Chem 6(18):3229
32. Shen J, Nguyen TT, Goh Y-P, Ye W, Fu X, Xu J, Tan C-H (2006) J Am Chem Soc 128:13692
33. Shen J, Tan CH (2008) Org Biomol Chem 6(22):4096
34. Jiang Z, Zhang Y, Ye W, Tan C-H (2007) Tetrahedron Lett 48:51
35. Fu X, Jiang Z, Tan CH (2007) Chem Commun 5058
36. Cho B, Tan C-H, Wong MW (2011) Org Biomol Chem 9:4550
37. Leow D, Shen J, Su Y, Peh G (2014) Mini Rev Org Chem 11:410
38. Thorat VH (2014) Synlett 25:1482
39. Leow D, Lin S, Chittimalla SK, Fu X, Tan C-H (2008) Angew Chem Int Ed 47:5641
40. Cho B, Tan C-H, Wong MW (2012) J Org Chem 77:6553
41. Lin S, Leow D, Huang K-W, Tan C-H (2009) Chem Asian J 4:1741
42. Leow D (2009) Chiral guanidine catalyzed enantioselective protonation reactions. Ph.D dissertation, National University of Singapore
43. Liu H, Leow D, Huang K-W, Tan C-H (2009) J Am Chem Soc 131:7212
44. Liu H, Feng W, Kee C-W, Leow D, Loh W-T, Tan C-H (2010) Adv Synth Catal 352:3373
45. Zhao Y, Lim X, Pan Y, Zong L, Feng W, Tan C-H, Huang K-W (2012) Chem Commun 48:5479
46. Wang J, Liu H, Fan Y, Yang Y, Jiang Z, Tan C-H (2010) Chem Eur J 16:12534
47. Wang J, Chen J, Kee CW, Tan C-H (2012) Angew Chem Int Ed 51:2382
48. Jiang Z, Yang YY, Pan YH, Zhao YJ, Liu HJ, Tan C-H (2009) Chem Eur J 15:4925
49. Jiang Z, Pan Y, Zhao Y, Ma T, Lee R, Yang Y, Huang K-W, Wong MW, Tan C-H (2009) Angew Chem Int Ed 48:3627
50. Jing Z, Liu J, Chin KF, Chen W, Tan C-H, Jiang Z (2014) Aust J Chem 67:1119
51. Pan Y, Zhao Y, Ma T, Yang Y, Liu H, Jiang Z, Tan C-H (2010) Chem Eur J 16:779
52. Zhao Y, Pan Y, Liu H, Yang Y, Jiang Z, Tan C-H (2011) Chem Eur J 17:3571
53. Zhao Y, Pan Y, Sim S-BD, Tan C-H (2012) Org Biomol Chem 10:479
54. Pan Y, Tan CH (2011) Synthesis 2044
55. Pan Y, Kee CW, Jiang Z, Ma T, Zhao Y, Yang Y, Xue H, Tan CH (2011) Chem Eur J 17:8363
56. Li L, Chen W, Yang W, Pan Y, Liu H, Tan C-H, Jiang Z (2012) Chem Commun 48:5124
57. Yang C, Chen W, Yang W, Zhu B, Yan L, Tan C-H, Jiang Z (2013) Chem Asian J 8:2960
58. Chen W, Yang W, Yan L, Tan C-H, Jiang Z (2013) Chem Commun 49:9854
59. Ishikawa T, Araki Y, Kumamoto T, Seki H, Fukuda K, Isobe T (2001) Chem Commun 245
60. Sohtome Y, Hashimoto Y, Nagasawa K (2005) Adv Synth Catal 347:1643
61. Terada M, Ube H, Yaguchi Y (2006) J Am Chem Soc 128:1454

62. Uyeda C, Jacobsen EN (2008) J Am Chem Soc 130:9228
63. Yu Z, Liu X, Zhou L, Lin L, Feng X (2009) Angew Chem Int Ed 48:5195
64. Misaki T, Takimoto G, Sugimura T (2010) J Am Chem Soc 132:6286
65. Zou L, Wang B, Mu H, Zhang H, Song Y, Qu J (2013) Org Lett 15:3106
66. Zhang Y, Kee CW, Lee R, Fu X, Soh JYT, Loh EMF, Huang K-W, Tan C-H (2011) Chem Commun 47:3897
67. Fu X, Loh W-T, Zhang Y, Chen T, Ma T, Liu H, Wang J, Tan C-H (2009) Angew Chem Int Ed 48:7387
68. Dong S, Liu X, Chen X, Mei F, Zhang Y, Gao B, Lin L, Feng X (2010) J Am Chem Soc 132:10650
69. Dong S, Liu X, Zhang Y, Lin L, Feng X (2011) Org Lett 13:5060
70. Xiao X, Liu X, Dong S, Cai Y, Lin L, Feng X (2012) Chem Eur J 18:15922
71. Dong S, Liu X, Zhu Y, He P, Lin L, Feng X (2013) J Am Chem Soc 135:10026
72. Ma T, Fu X, Kee CW, Zong L, Pan Y, Huang K-W, Tan C-H (2011) J Am Chem Soc 133:2828
73. Yang Y, Moinodeen F, Chin W, Ma T, Jiang Z, Tan C-H (2012) Org Lett 14:4762
74. Zong L, Ban X, Kee CW, Tan C-H (2014) Angew Chem Int Ed 53:11849
75. He W, Ge Y-C, Tan C-H (2014) Org Lett 16:3244

Top Heterocycl Chem (2017) 50: 157–178
DOI: 10.1007/7081_2015_170

Published online: 11 September 2015

# Conformationally Flexible Guanidine–(Thio)Urea Bifunctional Organocatalysts

**Kazuo Nagasawa, Minami Odagi, and Masaru Kato**

**Abstract** Enzymes can exhibit diverse catalytic functions by exploiting conformational changes in response to various external stimuli. An intriguing property of enzymes is that they can attain rate acceleration relative to competing reaction pathway occurring via other conformations. Inspired by these flexible but well-regulated functions of enzymes, we have developed the conformationally flexible guanidine-(thio)urea bifunctional organocatalyst. In this chapter, we describe some asymmetric reactions and very unique entropy-driven reactions catalyzed by guanidine-(thio)urea bifunctional organocatalyst.

**Keywords** Asymmetric reaction · Entropy-driven · Guanidine · Organocatalyst · Thiourea · Urea

## Contents



K. Nagasawa (✉), M. Odagi, and M. Kato
Tokyo University of Agriculture and Technology, Tokyo, Japan
e-mail: knaga@cc.tuat.ac.jp

# 1 Introduction

The functional groups of (thio)ureas, guanidiniums, and amidines have recently received much attention as catalytic core functional groups in organocatalysts, because they serve as effective double H-bond donors, which increase the coordination ability as well as reactivity of substrates [1–9]. With these interactions, organocatalysts bearing double H-bond donors control the asymmetric geometry of reaction components more efficiently compared to single H-bond donors. Furthermore, the characteristic properties of the functional groups of double H-bond donors can provide selective binding modes to specific substrates, which result in the formation of well-regulated transition-state architectures and thus afford high stereoselectivity in the bond-forming process.

We have recently developed bifunctional-type organocatalysts containing guanidine and (thio)urea functional groups with structurally flexible chiral backbones [10, 11]. With these catalysts, characteristic enantioselective reactions have been developed. In this chapter, we describe (1) the design concept for guanidine–(thio)urea bifunctional catalysts, (2) their utility in asymmetric reactions, and (3) applications to the natural product syntheses. We begin with our design concept for guanidine–(thio)urea bifunctional catalysts in Sect. 2. In Sect. 3, organocatalysts with diamine-tethered chiral spacers for 1,2-addition reactions such as the Henry reaction and hydroxylation of β-keto esters are described, as well as their application to natural product synthesis. In Sect. 4, we describe 1,3-diamine-tethered organocatalysts, which are considered to be more conformationally flexible than their 1,2-diamine-tethered counterparts, for Michael and related 1,4-addition reactions. We also cover an enantidivergent Mannich-type reaction (1,2-addition) and Friedel–Crafts-type reactions of nitroolefins (1,4-addition) in Sect. 4, where the roles of differential activation entropy and enthalpy are described.

# 2 Design of Guanidine- and (Thio)Urea-Containing Organocatalysts

The guanidine group, which is present in the side chain of the amino acid arginine, plays an important role in enzyme catalysis owing to its strong charge and hydrogen bonding-driven association with anionic species such as carboxylates, phosphates, and sulfates [9]. In contrast, hydrogen-bonding interactions with urea and thiourea groups

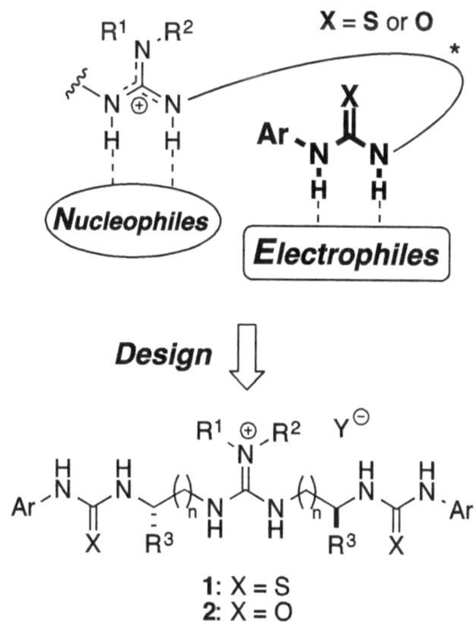


**Fig. 1** Design concept and structure of guanidine–(thio)urea bifunctional organocatalysts

are known to be effective in activating electrophiles [8]. These unique characteristics prompted us to develop a new type of catalyst framework that contains both guanidine- and thiourea-/urea-binding sites. The basic concept used in the design of these frameworks is shown in Fig. 1. We anticipated that catalysts bearing multiple double hydrogen-bonding donors, which act as Brønsted acids, would stabilize transition states in reactions of nucleophiles with electrophiles via a synergistic proximity effect. If the catalytic units of guanidine and (thio)urea groups are linked through a suitable chiral spacer, asymmetric induction is expected. In these catalysts, the frameworks can be easily tailored for compatibility with a variety of bond-forming reactions by tuning both the catalytic sites and chiral spacers. We also expected the flexibility of the chiral spacer to enable a rapid response to different transition-state architectures, which should thus allow for high rates of catalyst turnover. On the basis of these ideas, $C_2$-symmetrical guanidine-(thio)urea catalysts **1** and **2** were designed (Fig. 1). The catalysts, containing various chiral spacers ($R^3$) and the guanidinium substituents ($R^1$, $R^2$), can be easily synthesized starting from optically active amino acids.

# 3 Guanidine–(Thio)urea Bifunctional Catalysts for Asymmetric 1,2-Addition Reactions

## *3.1 Henry Reaction with Aldehydes*

The Henry reaction, nitroaldol reaction, is one of the fundamental carbon–carbon bond-forming reactions and proceeds between nitroalkanes and aldehydes under mild basic conditions. Since the resulting nitroalcohols provide synthetically useful chiral building blocks through simple and short transformations, many synthetic

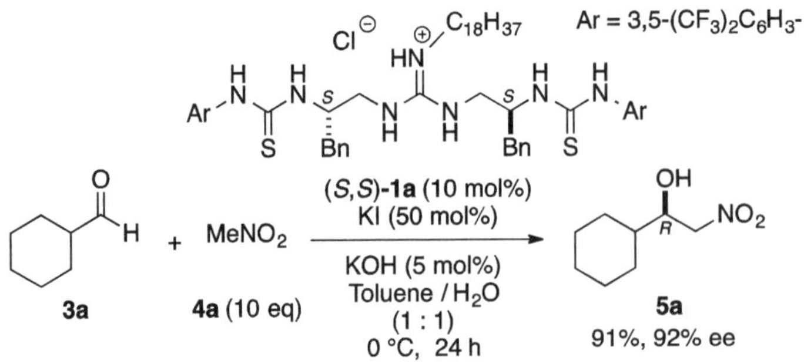

**Scheme 1** Enantioselective Henry reaction catalyzed by guanidine–thiourea catalyst (*S*,*S*)-**1a**

**Table 1** Time-course studies in the presence of various inorganic salts

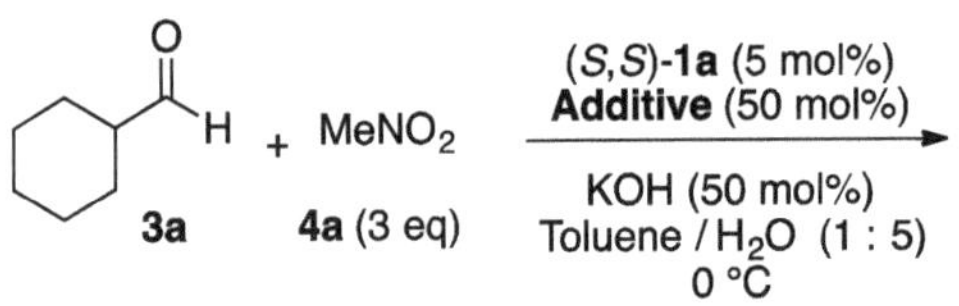

| Time | | 1 h | | 3 h | | 12 h | | 24 h | |
|---|---|---|---|---|---|---|---|---|---|
| | | Yield | *ee* | Yield | *ee* | Yield | *ee* | Yield | *ee* |
| Entry | Additive | (%) | (%) | (%) | (%) | (%) | (%) | (%) | (%) |
| 1 | None | 52 | 52 | 71 | 50 | 90 | 47 | 91 | 43 |
| 2 | $NaBF_4$ | 60 | 53 | 63 | 48 | 67 | 42 | 64 | 38 |
| 3 | $KBf_4$ | 55 | 61 | 60 | 59 | 75 | 47 | 77 | 47 |
| 4 | KBr | 58 | 72 | 62 | 70 | 63 | 63 | 76 | 57 |
| 5 | NaBr | 57 | 70 | 60 | 68 | 65 | 60 | 65 | 60 |
| 6 | NaI | 46 | 75 | 57 | 76 | 64 | 68 | 64 | 67 |
| 7 | KI | 66 | 75 | 74 | 75 | 88 | 74 | 88 | 74 |

efforts have been devoted to the use of organocatalysts as well as metal-containing catalysts [12–14].

Based on our bifunctional cooperative concept for guanidine–thiourea organocatalysts, we envisaged the guanidine recognizes the nucleophilic anion of nitronates while the thiourea simultaneously interacts with the aldehyde electrophile [15]. In the catalytic asymmetric Henry reaction, the retro-reaction should be suppressed to obtain high asymmetric induction. Thus, we installed a long, highly hydrophobic alkyl chain to the guanidine group, and the resulting catalyst (*S*,*S*)-**1a** was applied under biphasic conditions to avoid the undesired retro-process. As shown in Scheme 1, the Henry reaction product **5a** was obtained in high yield and high enantioselectivity by utilizing a catalytic amount of KOH (5 mol%) and KI (50 mol%) in toluene/$H_2O$ (1:1) at 0°C. In this reaction, the KI additive was found to play a significant role in suppressing the retro-process.

As shown in Table 1, the enantiomeric excess of the Henry reaction product **5a** was decreased as the reaction time was prolonged in the absence of any additives

**Table 2** Substrate scope for the catalytic asymmetric Henry reaction catalyzed by (*S*,*S*)-**1a**

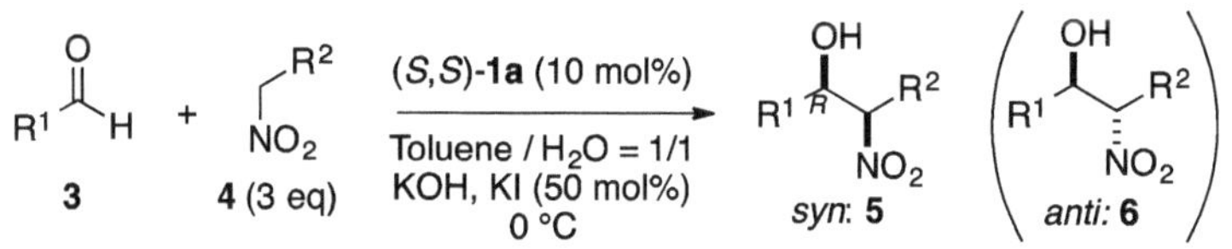


| Entry | 3 | ($R^1$) | 4 | ($R^2$) | KOH (mol%) | Time (h) | 5 Yield (%) | *ee* (%) | *Syn*/*anti* |
|---|---|---|---|---|---|---|---|---|---|
| 1 | **3b** | *c*-$C_5H_9$ | **4a** | H | 40 | 5 | 76 (**5b**) | 82 | – |
| 2[a] | **3c** | *t*-Bu | **4a** | H | 5 | 45 | 85 (**5c**) | 88 | – |
| 3[a] | **3d** | *i*-Pr | **4a** | H | 10 | 19 | 88 (**5d**) | 83 | – |
| 4 | **3e** | $Et_2CH$ | **4a** | H | 10 | 36 | 70 (**5e**) | 88 | – |
| 5 | **3f** | $Ph(CH_2)_2$ | **4a** | H | 5 | 18 | 79 (**5f**) | 55 | – |
| 6[a] | **3a** | *c*-$C_6H_{11}$ | **4b** | Me | 8 | 24 | 77 (**5g**) | 93 | 99/1 |
| 7[a] | **3d** | *i*-Pr | **4b** | Me | 8 | 24 | 50 (**5h**) | 90 | 97/3 |
| 8[a] | **3e** | $Et_2CH$ | **4b** | Me | 20 | 24 | 52 (**5i**) | 91 | 99/1 |
| 9[a] | **3g** | $Me_2CH(CH_2)_2$ | **4b** | Me | 5 | 24 | 58 (**5j**) | 99 | 93/7 |
| 10[a] | **3f** | $Ph(CH_2)_2$ | **4b** | Me | 10 | 24 | 76 (**5k**) | 83 | 90/10 |
| 11[a] | **3h** | *n*-Pr | **4b** | Me | 20 | 24 | 91 (**5l**) | 84 | 87/13 |
| 12[a] | **3i** | $TBSOCH_2$ | **4b** | Me | 6 | 48 | 63 (**5m**) | 98 | 86/14 |
| 13[a] | **3j** | $TBSO(CH_2)_2$ | **4b** | Me | 8 | 24 | 50 (**5n**) | 92 | 90/10 |
| 14[a] | **3a** | *c*-$C_6H_{11}$ | **4c** | Et | 5 | 40 | 61 (**5o**) | 95 | 99/1 |
| 15 | **3a** | *c*-$C_6H_{11}$ | **4d** | $CH_2OTBS$ | 7 | 48 | 63 (**5p**) | 90 | 99/1 |
| 16 | **3a** | *c*-$C_6H_{11}$ | **4e** | $CH_2OTIPS$ | 6 | 48 | 60 (**5q**) | 90 | 99/1 |
| 17 | **3a** | *c*-$C_6H_{11}$ | **4f** | Ph | 7 | 48 | 67 (**5r**) | 95 | 99/1 |
| 18[a] | **3h** | *n*-Pr | **4c** | Et | 5 | 48 | 63 (**5s**) | 85 | 90/10 |
| 19 | **3h** | *n*-Pr | **4d** | $CH_2OTBS$ | 3 | 48 | 51 (**5t**) | 87 | 93/7 |
| 20 | **3h** | *n*-Pr | **4e** | $CH_2OTIPS$ | 3 | 24 | 58 (**5u**) | 87 | 92/8 |
| 21 | **3h** | *n*-Pr | **4f** | Ph | 10 | 24 | 70 (**5v**) | 87 | 91/1 |

[a]10 equiv. of **4** was used.

(entry 1). Then, we investigated the effect of some inorganic salt additives in this reaction. With harder counter anions, as in $NaBF_4$ or $KBF_4$, only unsatisfactory results were obtained (entries 2 and 3). On the other hand, inorganic salts with softer anionic species were very effective (entries 4–7). Especially KI efficiently inhibited the retro-process of the reaction, and the Henry adduct **5a** was obtained in 88% yield with 74% *ee* after 24 h without any loss of the initial enantiomeric excess (entry 7). The KI dissociates the Henry product **5a** from catalyst **1a**, thereby completely suppressing the undesired **1a**-catalyzed retro-process. Accordingly, a highly enantioselective reaction was achieved.

The substrate scope for the **1a**-catalyzed enantioselective Henry reaction is shown in Table 2. Cyclic and linear α-branched aliphatic aldehydes **3b–e** afforded the corresponding nitroaldol products **5b–e** in good yields and with good *ee* (entries 1–4). In the case of unbranched aldehyde **3f**, a moderate selectivity was obtained (entry

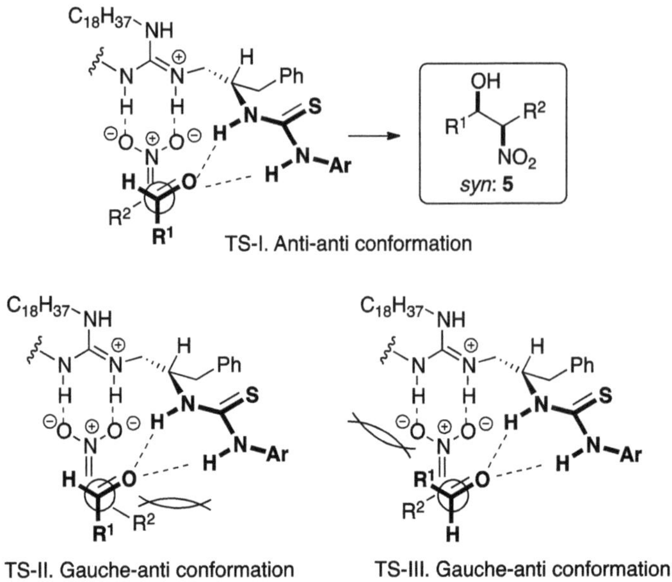


**Fig. 2** Plausible transition-state model for the enantioselective Henry reaction in the presence of **1a**

5). The diastereo- and enantioselective version of this Henry reaction was also examined with prochiral nitroalkanes **4b–f** (entries 6–21). Cyclic, α-branched, β-branched, unbranched, and heteroatom-substituted aldehydes **3a**, **d–j** all gave the corresponding Henry products **5g–n** with high diastereo- and enantioselectivities (entries 6–13). A variety of different nitroalkanes is tolerated, and excellent diastereo- and enantioselectivities were observed (entries 14–21). In all cases, the *syn*-adducts of **5** were obtained with high diastereoselectivity. Based on these results, the following transition state (TS) for the diastereo- and enantioselective Henry reaction catalyzed by **1a** was proposed (Fig. 2). Among the possible TS models, TS-1, where substituents in both aldehyde ($R^1$) and nitroalkanes ($R^2$) are orientated in an *anti* relationship (anti–anti conformation), is most favorable from the point of steric hindrance, which results in the *syn*-selective formation of product **5**.

## 3.2 Henry Reaction with α-Keto Esters

The construction of tetrasubstituted stereogenic centers, including tertiary alcohols via carbon–carbon bond formation, is a challenging task in synthetic chemistry. In an asymmetric Henry reaction of α-keto esters with nitroalkanes, highly functionalized chiral tertiary alcohols are formed [12–14]. We thus extended our guanidine–thiourea bifunctional catalyst concept for the Henry reaction of aldehydes with nitroalkanes (described in 3–1) to α-keto esters [16].

**Table 3** Catalytic asymmetric Henry reaction of α-keto esters **7** with nitroalkanes **3** catalyzed by (*S*,*S*)-**1a**

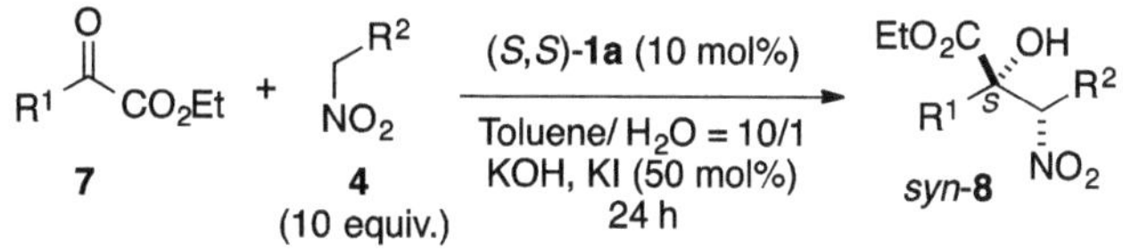


| Entry | 7 | ($R^1$) | 4 | ($R^2$) | KOH (mol%) | Temp. (°C) | 8 Yield (%) | ee (%) | Syn/anti |
|---|---|---|---|---|---|---|---|---|---|
| 1 | **7a** | *c*-$C_6H_{11}$ | **4a** | H | 10 | −20 | 83 (**8a**) | 80 | – |
| 2 | **7b** | *c*-$C_5H_9$ | **4a** | H | 10 | −25 | 90 (**8b**) | 93 | – |
| 3[a] | **7c** | *i*-Pr | **4a** | H | 5 | −30 | 62 (**8c**) | 80 | – |
| 4[a] | **7d** | $Et_2CH$ | **4a** | H | 10 | −25 | 89 (**8d**) | 78 | – |
| 5 | **7e** | *n*-Hex | **4a** | H | 10 | −25 | 60 (**8e**) | 83 | – |
| 6[b] | **7c** | *i*-Pr | **4b** | Me | 10 | −35 | 43 (**8f**) | 81 | 86/14 |
| 7[b] | **7c** | *i*-Pr | **4b** | Me | 10 | −35 | 45 (**8g**) | 80 | 85/15 |
| 8 | **7a** | *c*-$C_6H_{11}$ | **4b** | Me | 10 | −30 | 45 (**8h**) | 91 | 97/3 |
| 9 | **7a** | *c*-$C_6H_{11}$ | **4c** | Et | 10 | −30 | 35 (**8i**) | 83 | 82/8 |
| 10 | **7e** | *n*-Hex | **4b** | Me | 20 | −30 | 36 (**8j**) | 93 | 79/21 |

[a]Toluene/$H_2O$ = 5/1.
[b]**1a** (20 mol%), KI (100 mol%).

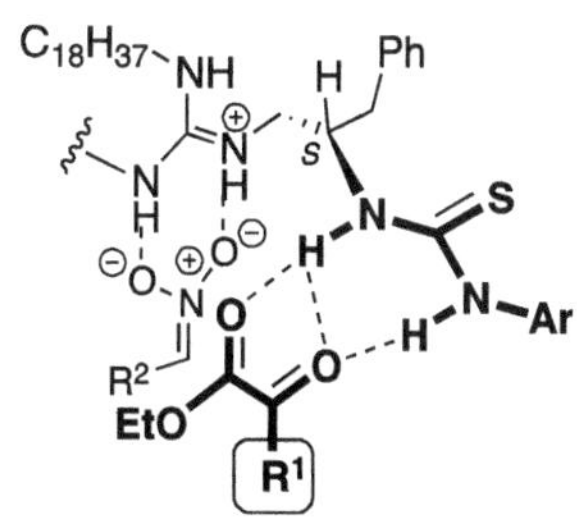


**Fig. 3** Transition-state model for the enantioselective Henry reaction with α-keto ester **7** and nitroalkane **4** in the presence of **1a**

In the case of Henry reactions with α-keto esters, the addition of a small amount of water to the toluene solvent and subzero temperature conditions were very effective. Under these conditions, the reaction proceeded on ice, and high enantioselectivity was obtained. Again, KI is mandatory for suppressing the retro-process of the reaction. A variety of aliphatic α-keto esters **7** and nitroalkanes **4** are suitable substrates under these conditions (Table 3).

Reactions of cyclic, branched, and linear aliphatic α-keto esters **7a–e** with nitromethane proceeded smoothly to give the corresponding Henry products **8a–e** with good enantioselectivities (entries 1–5). In the diastereo- and enantioselective version of this reaction, both high *syn*-selectivities and high enantioselectivities were obtained (entries 6–10). The stereochemistry of the products **8** is explained by the transition-state model shown in Fig. 3, which has *anti*-relationships between the $R^1$ group and the nitroalkane and the $R^2$ group and the α-keto ester, respectively, to minimize their steric repulsions (Fig. 3).

## 3.3 Aza-Henry Reaction

By changing the electrophile from aldehyde to imine in the Henry reaction, a new type of carbon–carbon bond-forming reaction, i.e., the aza-Henry reaction, would take place. This reaction provides synthetically useful β-nitro amines. Following our guanidine–thiourea bifunctional cooperative concept, an enantioselective aza-Henry reaction was expected to take place through the interaction between the catalyst thiourea and the imines electrophiles [17].

After screening the reaction conditions including variations of the catalyst structure, a highly enantioselective aza-Henry reaction was found to proceed with catalyst **1b** bearing a pyrrolidine on the guanidine and benzyl groups in the chiral spacers under solid–liquid biphasic conditions using $Cs_2CO_3$-THF at −10°C. In the aza-Henry reaction, the retro-process is not involved, and accordingly, no additives, such as KI in the Henry reaction, are required. Under optimized conditions, a variety of imines and nitroalkanes were investigated (Table 4, entries 1–14). The

**Table 4** Asymmetric aza-Henry reaction of imines **9** with nitroalkanes **10** in the presence of (*S*,*S*)-**1d**

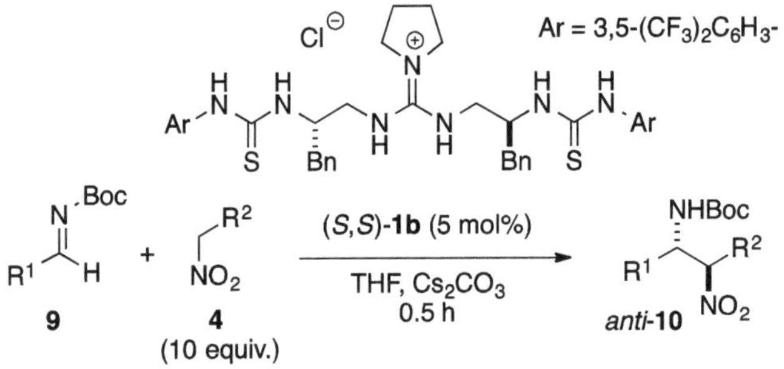


| | | | 9 | | 4 | | $Cs_2CO_3$ | Temp. | 10 |
|---|---|---|---|---|---|---|---|---|---|
| | | Entry | | | | ($R^1$) | | ($R^2$) | (mol%) |
| (°C) | | Yield (%) | | *Anti*/*syn* | *ee* (%) | | | | |
| 1 | **9a** | *c*-$H_6H_{11}$ | **4a** | H | 100 | 0 | 92 (**10a**) | 96 | – |
| 2 | **9b** | *c*-$H_5H_9$ | **4a** | H | 25 | −10 | 84 (**10b**) | 85 | – |
| 3 | **9c** | *i*-Pr | **4a** | H | 25 | −10 | 92 (**10c**) | 96 | – |
| 4 | **9d** | *n*-Hex | **4a** | H | 25 | 0 | 82 (**10d**) | 90 | – |
| 5 | **9e** | Ph | **4a** | H | 25 | −10 | 96 (**10e**) | 96 | – |
| 6 | **9f** | 4-Me-$C_6H_4$ | **4a** | H | 25 | −10 | 85 (**10f**) | 90 | – |
| 7 | **9g** | 4-Cl-$C_6H_4$ | **4a** | H | 25 | −10 | 88 (**10g**) | 96 | – |
| 8 | **9e** | Ph | **4b** | Me | 50 | −10 | 94 (**10h**) | 99 | 99/1 |
| 9 | **9e** | Ph | **4c** | Et | 50 | −20 | 90 (**10i**) | 99 | 99/1 |
| 10 | **9e** | Ph | **4d** | $CH_2OTBS$ | 50 | −20 | 85 (**10j**) | 98 | 94/6 |
| 11 | **9f** | 4-Me-$C_6H_4$ | **4b** | Me | 50 | −25 | 89 (**10k**) | 99 | 92/8 |
| 12 | **9f** | 4-Me-$C_6H_4$ | **4c** | Et | 50 | −25 | 89 (**10l**) | 97 | 90/10 |
| 13 | **9g** | 4-Cl-$C_6H_4$ | **4b** | Me | 25 | −10 | 81 (**10m**) | 97 | 96/4 |
| 14 | **9g** | 4-Cl-$C_6H_4$ | **4c** | Et | 25 | −10 | 94 (**10n**) | 96 | 95/5 |

aza-Henry reaction of nitromethane with cyclic, branched, linear, and aromatic mines **9a–g** proceeded well, and the corresponding adducts **10a–g** were obtained in high yields (82–96%) with high enantioselectivities (90–96% *ee*) (entries 1–7). In the corresponding diastereo- and enantioselective version with nitroalkanes **4b–d**, *anti*-β-nitro amines **10h–n** were obtained in good yields and with high diastereo- and enantioselectivities (90:10–99:1 *dr*, 96–99% *ee*) (entries 8–14).

## 3.4 Oxidation of Tetralone-Derived β-Keto Esters

Guanidine is known to interact with 1,3-dicarbonyl compounds through two parallel hydrogen bonds, while peroxides interact with urea groups. Thus, a guanidine–urea bifunctional catalyst was expected to promote the reaction of 1,3-dicarbonyl compounds with peroxides to generate α-hydroxy-β-1,3-dicarbonyl compounds in the presence of base. In this context, the enantioselective α-hydroxylation of β-keto esters derived from 1-tetralone was explored [18]. 1-Tetralones are challenging substrates for oxidation reactions because of their labile nature under oxidative conditions, and only limited examples of the α-hydroxylation to the corresponding α-hydroxy-β-keto esters have been reported with organocatalysts.

Using (*S*,*S*)-**2a** as the catalyst and cumene hydroperoxide (CHP) as the oxidant in the presence of $K_2CO_3$ in toluene, the α-hydroxylation of β-keto *tert*-butyl ester **11a** readily proceeded to give the hydroxylated product **12a** in 90% yield with 93% *ee* (Scheme 2). At a concentration of 0.05 M, any background oxidation reaction, i.e., reaction unmediated by the catalyst, was effectively suppressed, which improved the enantioselectivity.

This reaction was suitable for a variety of 1-tetralone-derived β-keto *tert*-butyl esters bearing substituents on the aromatic ring (Table 5). Thus, derivatives with both electron-donating (entries 1–3, 6) and electron-withdrawing (entries 4–5, 7–8) groups gave the corresponding hydroxylated compounds **12b–i** in high yields (82–99%) with high enantioselectivities (87–95% *ee*).

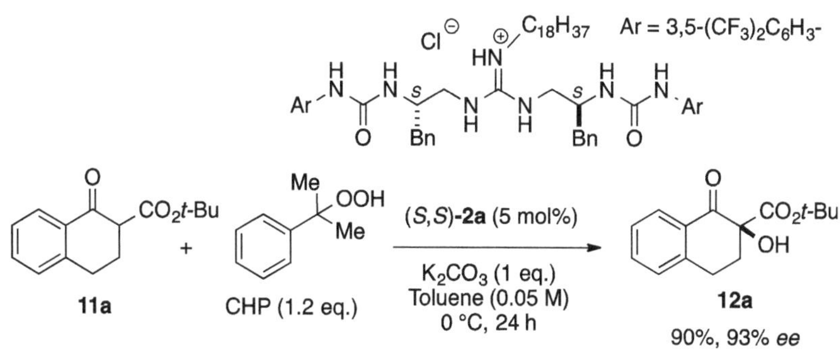


**Scheme 2** Asymmetric α-hydroxylation of β-keto ester **11a** catalyzed by guanidine–urea bifunctional catalyst (*S*,*S*)-**2a**

**Table 5** Asymmetric α-hydroxylation of tetralone-derived β-keto esters **11** catalyzed by (*S*,*S*)-**2a**

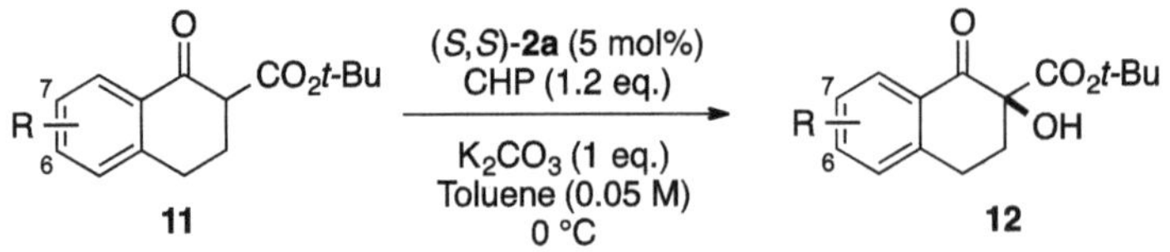


| | 11 | | Time | 12 | | |
|---|---|---|---|---|---|---|
| Entry | | (R) | (h) | | Yield (%) | *ee* (%) |
| 1 | **11b** | 6-OMe | 24 | **12b** | 91 | 87 |
| 2 | **11c** | 6-OBn | 16 | **12c** | 99 | 92 |
| 3 | **11d** | 6-$NMe_2$ | 48 | **12d** | 92 | 95 |
| 4 | **11e** | 6-Cl | 36 | **12e** | 88 | 95 |
| 5 | **11f** | 6-Br | 20 | **12f** | 83 | 85 |
| 6 | **11g** | 7-OMe | 16 | **12g** | 94 | 93 |
| 7 | **11h** | 7-Br | 20 | **12h** | 82 | 89 |
| 8 | **11i** | 7-F | 23 | **12i** | 90 | 90 |

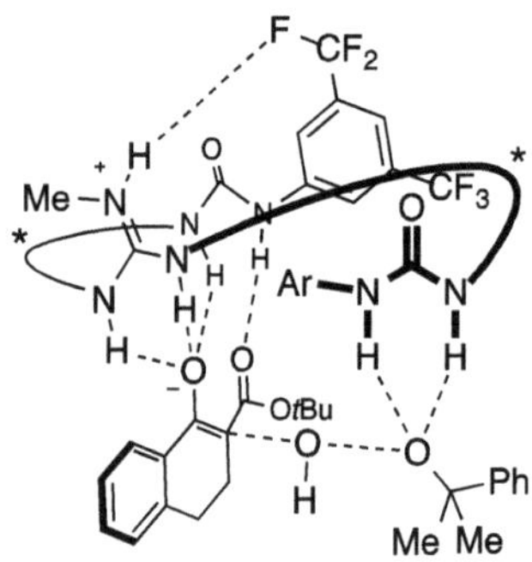


**Fig. 4** Proposed transition-state model (DFT calculation) for the α-hydroxylation of **11a** catalyzed by (*S*,*S*)-**2a**

The following transition-state model of the hydroxylation reaction catalyzed by **2a** was proposed based on DFT calculations (Fig. 4) [19]. In this model, four NH residues of the guanidinium and one urea group coordinate to two carbonyl groups in the keto ester substrate, effectively controlling the orientation of the keto ester enolate. The remaining two NH groups of the second urea group are suggested to activate cumene hydroperoxide through hydrogen bonding with the oxygen attached to the carbon in cumene. Through these multiple interactions, high enantioselectivity is suggested to arise.

The presented catalytic asymmetric hydroxylation was applied to a more complex substrate **13**, bearing a vinylogous β-keto moiety in the 2-pyridone structure. The reaction of **13** with CHP in the presence of guanidine–urea catalyst (*S*,*S*)-**2b** gave hydroxylated product **14** in 93% yield with 84% *ee* (Scheme 3) [20]. The enantiomeric excess of **14** was increased to 93% *ee* by a single recrystallization. The obtained product **14** is well known as a key synthetic intermediate in the

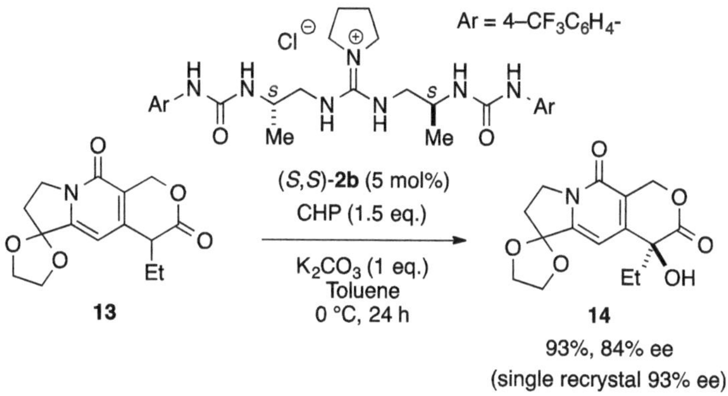


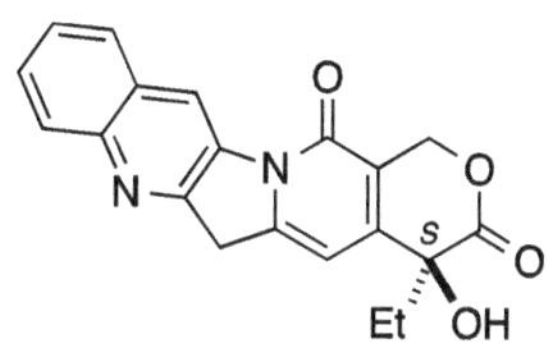
(20*S*)-Camptothecin (**15**)

**Scheme 3** Asymmetric α-hydroxylation of a lactone **13** with a vinylogous β-keto ester moiety catalyzed by (*S*,*S*)-**2b**

synthesis of (20*S*)-camptothecin (**15**), which shows significant antitumor activity through inhibition of type I topoisomerase (Topo I). This synthetic strategy allows us the synthesis of different camptothecin derivatives by simple variation of the substituent at C20.

## 3.5 *Oxidative Kinetic Resolution of Tetralone-Derived β-Keto esters*

β- and γ-Substituted tetralones are ubiquitous in many biologically active molecules; however, derivatization at the β- and γ-position is difficult, since the corresponding precursors for Michael reactions, i.e., α,β-unsaturated tetralones, are quite unstable and quickly aromatized [21, 22]. In this context, we envisaged to obtain chiral β- and γ-substituted tetralones via an oxidative kinetic resolution process [19].

In the oxidative kinetic resolution reaction for substituted tetralone-derived keto ester, guanidine–urea bifunctional catalyst **2c** bearing phenyl groups on the chiral spacers is effective. Moreover, high dilution conditions of 0.03 M and 0.75 equiv. of oxidant of CHP are mandatory. Under these conditions, a highly enantioselective oxidative kinetic resolution reaction proceeded with wide range of substrates

**Table 6** Oxidative kinetic resolution of β- or γ-substituted tetralone-derived keto esters *rac*-**16**

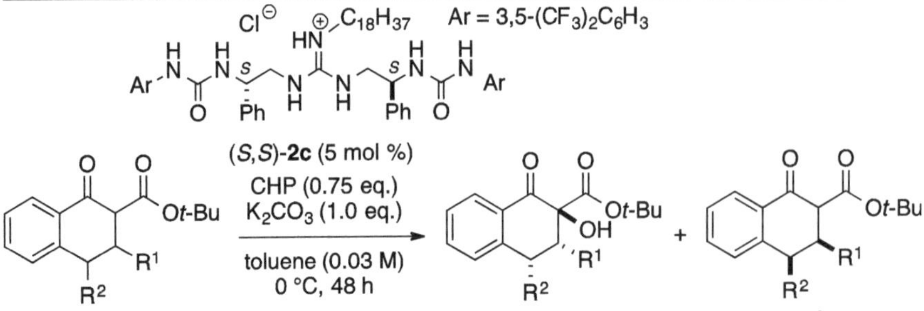


| Entry | *rac*-**16** | | | **17** | | **16** | | *s* value |
|---|---|---|---|---|---|---|---|---|
| | | ($R^1$) | ($R^2$) | Yield (%) | *ee* (%) | Yield (%) | *ee* (%) | |
| 1 | *rac*-**16a** | Ph | H | 49 (**17a**) | 83 | 42 (**16a**) | 97 | 44 |
| 2 | *rac*-**16b** | 2-Cl-$C_6H_4$ | H | 50 (**17b**) | 90 | 44 (**16b**) | 94 | 43 |
| 3 | *rac*-**16c** | 2-$CF_3$-$C_6H_4$ | H | 50 (**17c**) | 90 | 43 (**16c**) | 99 | 99 |
| 4 | *rac*-**16d** | 3-OMe-$C_6H_4$ | H | 48 (**17d**) | 89 | 44 (**16d**) | 89 | 51 |
| 5 | *rac*-**16e** | 3,5-OMe-$C_6H_4$ | H | 42 (**17e**) | 89 | 44 (**16e**) | 90 | 52 |
| 6 | *rac*-**16f** | Me | H | 43 (**17f**) | 90 | 43 (**16f**) | 67 | 38 |
| 7 | *rac*-**16g** | $CH_2CH_2Ph$ | H | 45 (**17g**) | 88 | 42 (**16g**) | 74 | 34 |
| 8 | *rac*-**16h** | H | Ph | 51 (**17h**) | 89 | 44 (**16h**) | 99 | 89 |
| 9 | *rac*-**16i** | H | Me | 48 (**17i**) | 91 | 45 (**16i**) | 92 | 69 |

(Table 6). In the case of β-substituted tetralones *rac*-**16a–e**, electron-donating and electron-withdrawing groups on the phenyl group at the β-position were well tolerated (entries 1–5), and high enantioselectivities were obtained for both the oxidation products **17a–e** (83–90% *ee*) and the recovered starting materials **16a–e** (89–97% *ee*). High *s* values of 43–99 were obtained. With compounds *rac*-**16f** and **16g,** bearing methyl and phenethyl substituents at the β-position, high selectivities for the oxidized products **17f** and **17g** (90 and 88% *ee*) and moderate to good selectivities for recovered **16f** and **16g** (67 and 74% *ee*) were obtained (entries 6 and 7). Oxidative kinetic resolutions also proceeded efficiently with γ-substituted tetralones *rac*-**16h** and **16i** (entries 8 and 9). High *s* values of 89 and 69 were obtained for these substrates, respectively.

The oxidative kinetic resolution of tetralones provides a new and efficient synthetic strategy for the synthesis of natural products. (+)-Linoxepin (**18**) is a caffeic acid dimer isolated from *Linum perenne* [23]. In this natural product a β-substituted tetralone moiety is embedded. Oxidative kinetic resolution of *rac*-**16j** proceeded with (*R*,*R*)-**2c** and CHP (0.75 equiv.) in toluene, and (−)-**16j** was obtained in 37% yield with 99% *ee* (Scheme 4). The optically active tetralone derivative (−)-**16j** was then efficiently converted into (+)-linoxepin (**18**) in 5 steps.

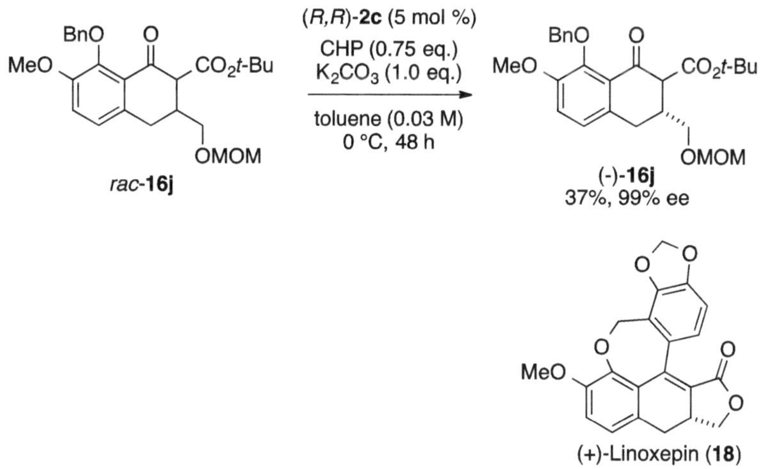


**Scheme 4** Synthesis of (+)-linoxepin (**18**) based on the oxidative kinetic resolution strategy for *rac*-**16j**

# 4 Guanidine–Thiourea Bifunctional Catalysts for Asymmetric 1,4-Addition Reactions

## *4.1 Michael Reaction of Nitroolefins*

The Michael reaction of β-dicarbonyl compounds with unsaturated electrophiles is a typical 1,4-type addition reaction and well recognized as a powerful carbon–carbon bond-forming reaction [24–26]. In the reaction with nitroolefins as unsaturated electrophiles, synthetically useful γ-nitrocarbonyls for the synthesis of chiral pyrrolidines and γ-amino acids are generated. Since the β-dicarbonyl compounds coordinate with guanidine in their enolate form, while nitroolefins interact with the thiourea group, bifunctional guanidine–thiourea organocatalysts were applied for this enantioselective version of the Michael reaction [27].

The Michael reaction of acetylacetone (**19a**) with nitrostyrene (**20a**) was explored using alanine-derived 1,2-diamine- and 1,3-diamine-tethered guanidine–thiourea bifunctional catalysts **1c** or **1d** in the presence of $K_2CO_3$ in toluene (Scheme 5). In the case of 1,2-diamine-tethered catalyst **1c**, the Michael adduct **21a** was obtained in 27% yield with 2% *ee*. On the other hand, adduct **21a** was obtained in 81% yield with 93% *ee* in the presence of the 1,3-diamine-tethered catalyst **1d**. These results suggest that for the 1,4-type addition reaction, a longer chiral spacer would be required as compared to the previously presented 1,2-type addition reactions, in order to preorganize the position of the nucleophile and electrophiles through the interaction with the guanidine and the thiourea, respectively.

The scope for β-dicarbonyls **19** with nitrostyrene (**20a**) in the Michael reaction is shown in Table 7. Acyclic 1,3-diketones **19a–c**, β-keto ester **19d**, and β-keto amid

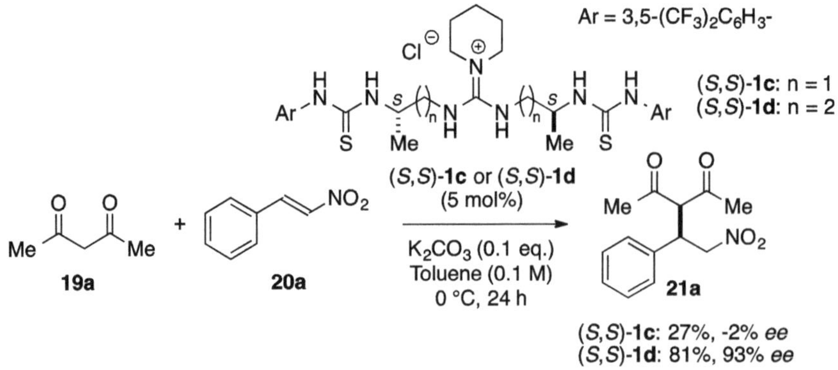


**Scheme 5** Asymmetric Michael reaction of acetylacetone (**19a**) with nitroalkene (**20a**) catalyzed by 1,2-diamine- and 1,3-diamine-tethered guanidine–thiourea bifunctional catalysts **1c** and **1d**

**Table 7** Asymmetric Michael reaction of β-dicarbonyls **19** with nitrostyrene (**20a**) catalyzed by 1,3-diamine-tethered catalyst **1d**

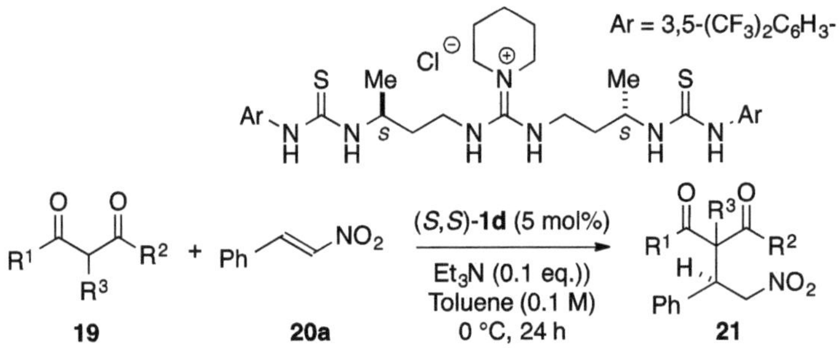


| Entry | 19 | | 21 | | | dr |
|---|---|---|---|---|---|---|
| | | | | Yield (%) | *ee* (%) | |
| 1 | **19a** | 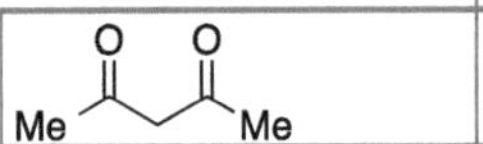 | **21a** | 99 | 93 | – |
| 2 | **19b** | 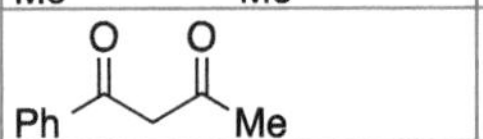 | **21b** | 88 | 94 | 4:1 |
| 3 | **19c** | 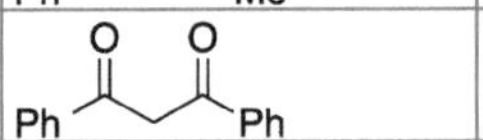 | **21c** | 65 | 90 | – |
| 4 | **19d** | 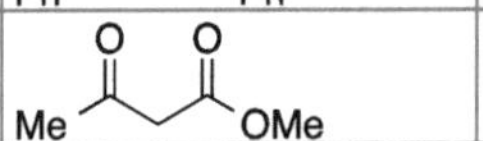 | **21d** | 98 | 92 | 1.8:1 |
| 5 | **19e** | 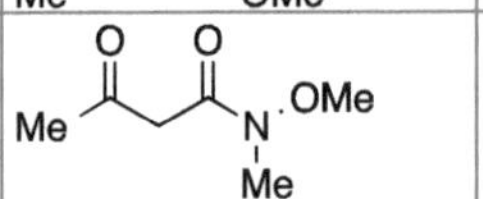 | **21e** | 96 | 93 | 15:1 |
| 6 | **19f** | 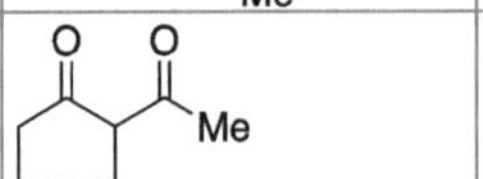 | **21f** | 98 | 91 | 16:1 |
| 7 | **19g** | 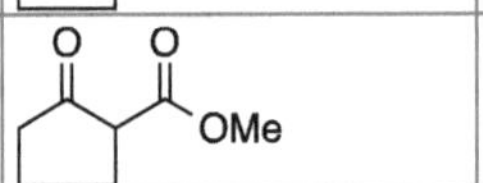 | **21g** | 99 | 91 | 99:1 |
| 8 | **19h** | 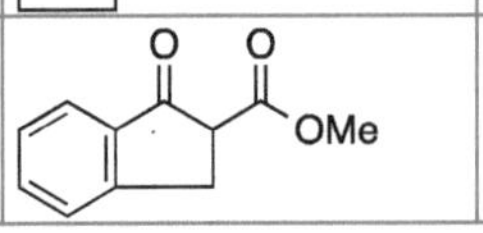 | **21h** | 98 | 80 | 3:1 |

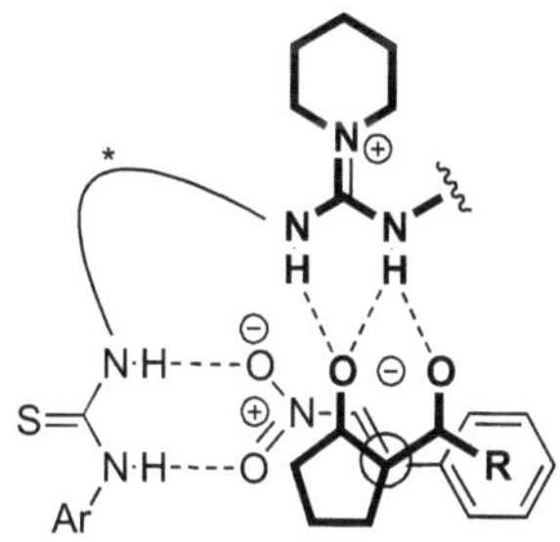


**Fig. 5** Proposed transition-state model for the Michael reaction of β-keto ester with nitrostyrene (**20a**) catalyzed by **1d**

**19e** are all suitable substrates (entries 1–5), and the corresponding Michael adducts **19a–e** were obtained in good to high yields (65–99%) with high enantioselectivities (90–94% *ee*). In the case of the cyclic 1,3-diketone **19f** and β-keto esters **19g** and **19h**, the reaction also proceeded effectively in high yields and with high enantioselectives (entries 6–8). In some cases, high diastereoselectivities were obtained as well (entries 5–7). The diastereoselection in entry 7 can be explained as follows (Fig. 5). The five-membered ring is favored to be at the nitro group site in nitrostyrene to avoid steric repulsion with the phenyl group, and accordingly, a (2*R*,3*S*)-stereochemistry was generated in the major isomer.

## 4.2 Phospha-Michael Reaction of Nitroolefins

Based on the cooperative concept in our guanidine–thiourea bifunctional organocatalysts, phosphonate is another good candidate for the nucleophile since specific interaction is known to occur with guanidines through phosphite formation. Thus, the asymmetric phospha-Michael reaction was realized with phosphonates and a nitroalkene in the presence of a guanidine–thiourea catalyst [28]. This reaction provides valuable precursors for biologically important β-aminophosphoric acids [29].

The phospha-Michael reaction of diphenyl phosphonate (**22**) with β-nitrostyrene (**20a**) proceeded in the presence of 1 mol% of catalyst **1e** and $K_2CO_3$ (0.5 equiv.) in a toluene/$H_2O$-mixed solvent system, and phospha-Michael adduct **23a** was obtained in 99% yield with 95% *ee* (Scheme 6). This catalytic reaction proceeded on a 5 mmol scale without any loss of reactivity or enantioselectivity.

The scope of the catalytic phospha-Michael reaction with a range of nitroalkenes is shown in Table 8. *para*-Substituted aromatic nitroalkenes including electron-withdrawing and electron-donating groups gave the corresponding phospha-Michael adducts in high yields (88–90%) and with high enantioselectivities (90–96% *ee*) (entries 1–3). Naphthyl-, 2-thienyl-, and 2-furyl-substituted nitroalkenes are also tolerated, and adducts were obtained in high yields with high enantioselectivities (entries 4–7). It is noteworthy that extraordinary high reactivity and enantioselectivity were observed for sterically bulky aliphatic nitroalkene **21i**, resulting in 95% yield and 92% *ee* (entry 8).

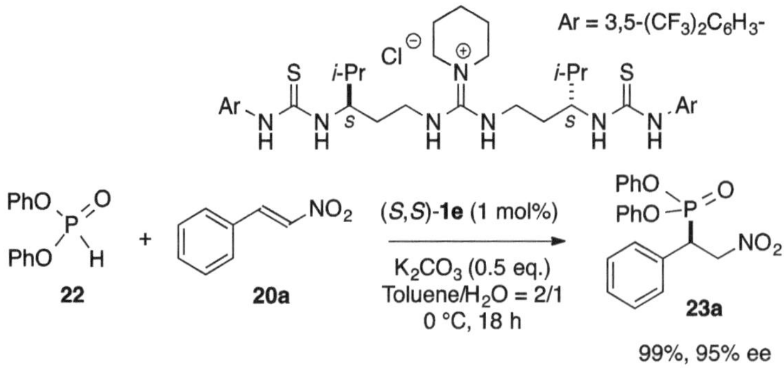

**Scheme 6** Asymmetric phospha-Michael reaction of diphenyl phosphonate (**22**) with β-nitrostyrene (**20a**) in the presence of **1e**

**Table 8** Catalytic phospha-Michael reaction of diphenyl phosphonate (**22**) with β-nitroalkenes using **1e**

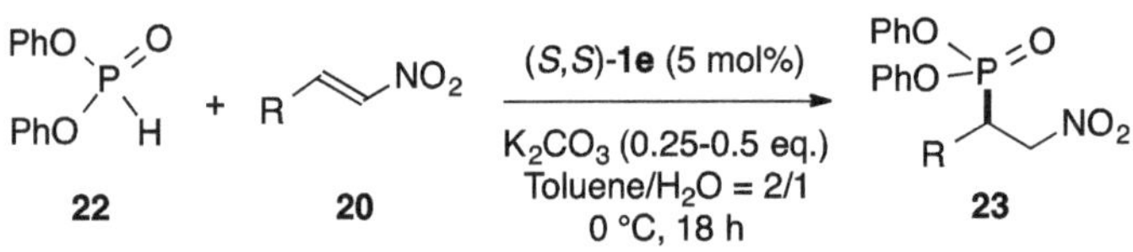

| | **20** | | **23** | | |
|---|---|---|---|---|---|
| Entry | | R | | Yield (%) | *ee* (%) |
| 1 | **20b** | 4-Me-$C_6H_4$ | **23b** | 88 | 91 |
| 2 | **20c** | 4-Cl-$C_6H_4$ | **23c** | 88 | 90 |
| 3 | **20d** | 4-MeO-$C_6H_4$ | **23d** | 90 | 96 |
| 4 | **20e** | 1-Naphthyl | **23e** | 84 | 96 |
| 5 | **20f** | 2-Naphthyl | **23f** | 81 | 98 |
| 6 | **20g** | 2-Thienyl | **23g** | 77 | 95 |
| 7 | **20h** | 2-Furyl | **23h** | 78 | 95 |
| 8 | **20i** | *t*-Bu | **23i** | 95 | 92 |

# 5 Entropy-Related Reactions Catalyzed by Conformationally Flexible Guanidine–Thiourea Bifunctional Compounds

## *5.1 Solvent-Dependent Enantiodivergent Mannich-Type Reaction (1,2-Addition)*

Enantiodivergent synthesis utilizing a single catalyst is one of the most efficient strategies for obtaining both enantiomers of a given product [30, 31]. Conformationally flexible guanidine–thiourea compounds allow the construction of enantiomerically distinct chiral environments with a single chiral source in

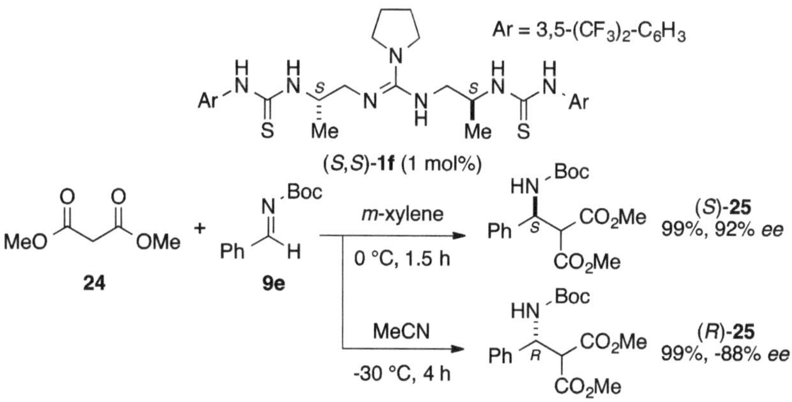


**Scheme 7** Solvent-dependent enantiodivergent Mannich-type reaction of methyl malonate (**24**) with aldimine **9e** catalyzed by **1f**

response to various external stimuli, i.e., reaction conditions. Thus, the relative arrangement of guanidine and thiourea in the catalyst can be selectively altered through conformational change in the flexible acyclic chiral spacer depending upon the reaction conditions (external stimuli), and accordingly, enantiodivegent reactions can be realized.

In this context, the Mannich-type reaction (1,2-addition) of *N*-Boc aldimine **9e** with methyl malonate (**24**) was explored as enantiodivergent reaction using conformationally flexible guanidine–thiourea organocatalyst [32]. After extensive screening, we found that catalyst **1f** shows characteristic solvent-dependent enantiodivergence in the Mannich-type reaction (Scheme 7). Thus, in nonpolar solvents such as toluene or *m*-xylene, Mannich adduct (*S*)-**25** was obtained in 99% yield with 92% *ee*. In contrast, the (*R*)-**25** adduct was observed in the polar aprotic solvent acetonitrile in 99% yield with 88% *ee*. It is noteworthy that the reaction rate enabled by catalyst **1f** is quite high, and turnover frequencies (TOF) of 66 $h^{-1}$ for the *S*-selective reaction and 25 $h^{-1}$ for the *R*-selective reaction were reached in the presence of 1 mol% of catalyst.

The kinetic analysis of the solvent-dependent enantiodivergent Mannich-type reaction by Eyring analysis revealed that the stereoselectivites in the *S*-selective Mannich-type reaction in nonpolar solvents are governed by the differences in the entropies of activation ($\Delta\Delta S^{\ddagger}_{S-R}$), whereas the stereodiscrimination processes in the *R*-selective reactions are governed by the differences in the enthalpies of activation ($\Delta\Delta H^{\ddagger}_{R-S}$).

## 5.2 *Friedel–Crafts-Type Reactions of Phenols with Nitroolefins (1,4-Addition)*

The 1,3-diamine-tethered guanidine–thiourea bifunctional catalysts, i.e., catalysts incorporating a longer chiral spacer, should display more dynamic motions

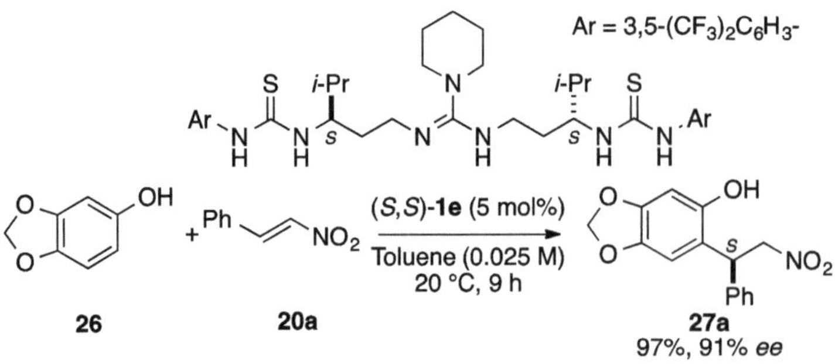

Scheme 8 Asymmetric FC-type reaction of sesamol (**26**) with nitrostyrene (**20a**) catalyzed by **1e**

**Table 9** Asymmetric FC reaction of sesamol (**26**) with nitroolefins **20** catalyzed by **1e**

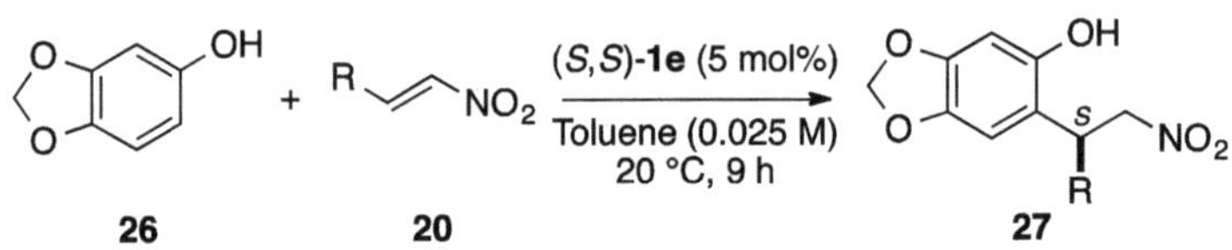

| Entry | 20 | | Time | 27 | | |
|---|---|---|---|---|---|---|
| | | (R) | (h) | | Yield (%) | *ee* (%) |
| 1 | **20c** | 4-Cl-$C_6H_4$ | 12 | **27b** | 99 | 90 |
| 2 | **20d** | 4-MeO-$C_6H_4$ | 12 | **27c** | 91 | 87 |
| 3 | **20f** | 2-Naphthyl | 12 | **27d** | 99 | 94 |
| 4 | **20g** | 2-Thienyl | 9 | **27e** | 98 | 90 |
| 5 | **20h** | 2-Furyl | 12 | **27f** | 99 | 88 |
| 6 | **20j** | *c*-Hex | 48 | **27g** | 84 | 91 |
| 7 | **20k** | *n*-Hex | 24 | **27h** | 66 | 90 |
| 8 | **20l** | Ph$(CH_2)_2$ | 24 | **27i** | 88 | 89 |

compared to the 1,2-diamine-tethered catalysts. This increased flexibility is expected to lead to an increase in the magnitude of the differential activation entropy ($\Delta\Delta S^{\ddagger}_{S-R}$) associated with the stereodiscrimination. This section deals with the entropy-related 1,4-addition reaction of phenols to nitroolefins (1,4-type FC reaction) catalyzed by a 1,3-diamine-tethered guanidine–thiourea bifunctional catalyst [33].

Catalyst **1e** with a 1,3-diamine tether promotes the *ortho*-selective 1,4-type FC alkylation reaction of sesamol (**26**) to nitrostylene (**20a**), and the FC-type adduct **27a** was obtained in 97% yield with 91% *ee* under high dilution conditions of 0.025 M in toluene at 20°C (Scheme 8).

The substrate scope for this reaction is shown in Table 9. Nitroolefins with aromatic groups (entries 1–5) as well as aliphatic substituents (entries 6–8) gave the corresponding Friedel–Crafts products **27b–i** in high yields (66–99%) with high enantioselectivities (87–94% *ee*).

**Table 10** Temperature dependence of the enantioselectivity in the **1e**-catalyzed enantioselective FC reaction of **26** with **20a** at various concentrations

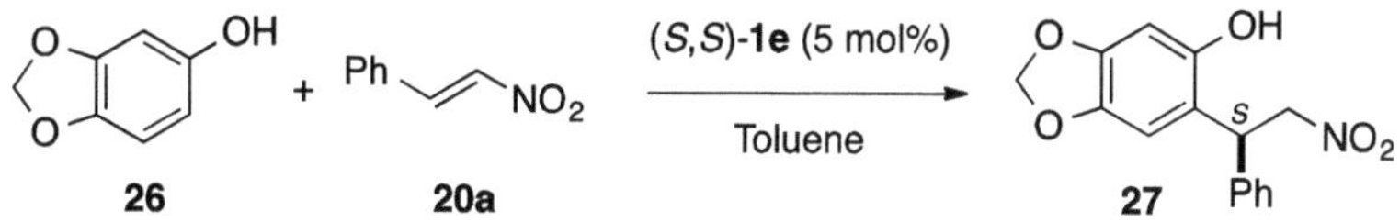


| Entry | Temp. (°C) | Conc. (M) | Time (h) | 27 Yield (%) | *ee* (%) |
|---|---|---|---|---|---|
| 1 | −30 | 0.1 | 24 | 65 | 69 |
| 2 | −20 | 0.1 | 18 | 88 | 75 |
| 3 | −10 | 0.1 | 6 | 88 | 78 |
| 4 | 0 | 0.1 | 6 | 89 | 79 |
| 5 | 10 | 0.1 | 6 | 93 | 80 |
| 6 | 20 | 0.1 | 2 | 96 | 85 |
| 7 | −30 | 0.05 | 24 | 66 | 80 |
| 8 | −20 | 0.05 | 18 | 87 | 83 |
| 9 | −10 | 0.05 | 9 | 88 | 85 |
| 10 | 0 | 0.05 | 9 | 89 | 86 |
| 11 | 10 | 0.05 | 9 | 96 | 87 |
| 12 | 20 | 0.05 | 9 | 97 | 89 |
| 13 | −30 | 0.025 | 24 | 66 | 91 |
| 14 | −20 | 0.025 | 18 | 83 | 91 |
| 15 | −10 | 0.025 | 9 | 88 | 91 |
| 16 | 0 | 0.025 | 9 | 90 | 91 |
| 17 | 10 | 0.025 | 9 | 96 | 91 |
| 18 | 20 | 0.025 | 9 | 97 | 91 |

This catalytic FC reaction showed very unique temperature dependency profiles (Table 10). Thus, the enantioselectivity increased as the reaction temperature increased at concentrations of 0.1 M and 0.05 M (entries 1–6, and 7–12). On the other hand, a constant enantioselectivity of 91% *ee* was obtained in the temperature range between −30 and 20°C at a concentration of 0.025 M (entries 13–18). The kinetic analysis for the **1e** catalyzed FC reaction was carried out based on an Erying equation as shown in Eq. (1):

$$\ln(k_S/k_R) = -\Delta\Delta H^{\ddagger}{}_{S-R}/RT + \Delta\Delta S^{\ddagger}{}_{S-R}/R \qquad (1)$$

According to Eq. (1), Erying plots of the **1e**-catalyzed FC reaction were obtained as shown in Fig. 6. In these plots, negative slopes and positive *y*-intercepts are obtained at less than threshold concentration. These results suggest that differential activation entropy ($\Delta\Delta S^{\ddagger}{}_{S\text{–}R}$) contributes to lowering the $\Delta\Delta G^{\ddagger}{}_{S\text{–}R}$ ($=\Delta\Delta H^{\ddagger}{}_{S-R} - \mathrm{T}\Delta\Delta S^{\ddagger}{}_{S\text{–}R}$), with an unfavorable enthalpic contribution ($\Delta\Delta H^{\ddagger}{}_{S\text{–}R}$). Quite interestingly, the value of $\Delta\Delta H^{\ddagger}{}_{S\text{–}R}$ reaches to zero at a concentration of

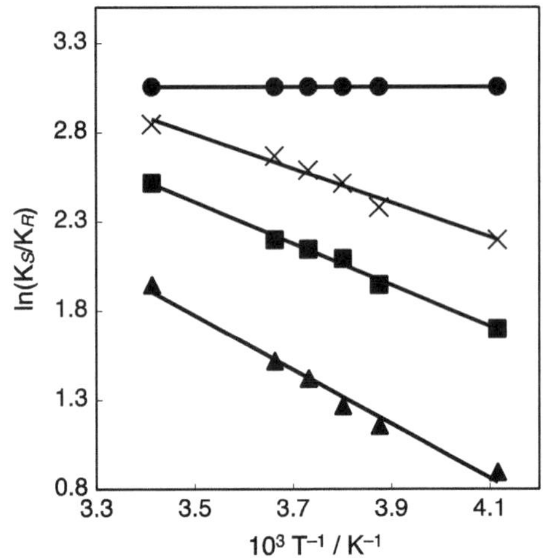


**Fig. 6** Eyring plots for temperature dependence **1e**-catalyzed enantioselective FC reaction of **26** with **20a** at various concentrations: 0.025 M (*circle*), 0.05 M (*cross*), 0.1 M (*square*), and 0.2 M (*triangle*)

0.025 M, and $\Delta\Delta S^{\ddagger}_{S\text{–}R}$ becomes responsible for stereodiscrimination. Thus, the activation entropy differences were identified to play a principal role in the stereodiscrimination process in the **1e**-catalyzed FC reaction. The present entropy-controlled catalytic reaction has the advantage that it does not require fine-tuning of the reaction temperature to attain maximum enantioselectivity.

## 6 Summary

In this chapter, we overviewed our guanidine–(thio)urea bifunctional organocatalysts from the design concept to a variety of enantioselective reactions together with the construction of chiral transition-state environments. We believe the concepts of flexible structures in catalysts will open new ideas for further catalysts design. The combination of weak H-bonding interactions with a highly flexible structure in these chiral catalysts effectively provides high selectivity together with high catalytic activity. The chiral environments constructed by the flexible catalyst structures can be easily tuned by altering the reaction conditions. The dynamic structural response of the organocatalyst provides unique stereoselection processes, including retro-free Henry reactions and enantiodivergent Mannich-type reactions, which are governed by entropy factors. We believe our approach will lead to the development of a range of stimulus-responsive smart catalysts that enable the construction of useful chiral molecules in an efficient and stereodivergent manner.

## References

1. Schreiner PR (2003) Metal-free organocatalysis through explicit hydrogen bonding interactions. Chem Soc Rev 32(5):289–296. doi:10.1039/B107298f

2. Takemoto Y (2005) Recognition and activation by ureas and thioureas: stereoselective reactions using ureas and thioureas as hydrogen-bonding donors. Org Biomol Chem 3 (24):4299–4306. doi:10.1039/b511216h
3. Akiyama T, Itoh J, Fuchibe K (2006) Recent progress in chiral brønsted acid catalysis. Adv Synth Catal 348(9):999–1010. doi:10.1002/adsc.200606074
4. Connon SJ (2006) Organocatalysis mediated by (thio)urea derivatives. Chemistry 12 (21):5418–5427. doi:10.1002/chem.200501076
5. Taylor MS, Jacobsen EN (2006) Asymmetric catalysis by chiral hydrogen-bond donors. Angew Chem Int Ed 45(10):1520–1543. doi:10.1002/anie.200503132
6. Doyle AG, Jacobsen EN (2007) Small-molecule H-bond donors in asymmetric catalysis. Chem Rev 107(12):5713–5743. doi:10.1021/cr068373r
7. Palomo C, Oiarbide M, Lopez R (2009) Asymmetric organocatalysis by chiral Bronsted bases: implications and applications. Chem Soc Rev 38(2):632–653. doi:10.1039/b708453f
8. Connon S (2009) The design of novel, synthetically useful (thio)urea-based organocatalysts. Synlett 2009(03):0354–0376. doi:10.1055/s-0028-1087557
9. Leow D, Tan CH (2009) Chiral guanidine catalyzed enantioselective reactions. Chem Asian J 4(4):488–507. doi:10.1002/asia.200800361
10. Nagasawa K, Sohtome Y (2009) The design of chiral double hydrogen bonding networks and their applications to catalytic asymmetric carbon-carbon and carbon-oxygen bond-forming reactions. Synlett 2010(01):1–22. doi:10.1055/s-0029-1218542
11. Sohtome Y, Nagasawa K (2012) Dynamic asymmetric organocatalysis: cooperative effects of weak interactions and conformational flexibility in asymmetric organocatalysts. Chem Commun 48(63):7777–7789. doi:10.1039/c2cc31846f
12. Boruwa J, Gogoi N, Saikia PP, Barua NC (2006) Catalytic asymmetric Henry reaction. Tetrahedron Asymmetry 17(24):3315–3326. doi:10.1016/j.tetasy.2006.12.005
13. Palomo C, Oiarbide M, Laso A (2007) Recent advances in the catalytic asymmetric nitroaldol (Henry) reaction. Eur J Org Chem 2007(16):2561–2574. doi:10.1002/ejoc.200700021
14. Alvarez-Casao Y, Marques-Lopez E, Herrera RP (2011) Organocatalytic enantioselective Henry reactions. Symmetry 3(4):220–245. doi:10.3390/sym3020220
15. Sohtome Y, Takemura N, Takada K, Takagi R, Iguchi T, Nagasawa K (2007) Organocatalytic asymmetric nitroaldol reaction: cooperative effects of guanidine and thiourea functional groups. Chem Asian J 2(9):1150–1160. doi:10.1002/asia.200700145
16. Takada K, Takemura N, Cho K, Sohtome Y, Nagasawa K (2008) Asymmetric organocatalytic nitroaldol reaction of α-keto esters: stereoselective construction of chiral tertiary alcohols at subzero temperature. Tetrahedron Lett 49(10):1623–1626. doi:10.1016/j.tetlet.2008.01.030
17. Takada K, Nagasawa K (2009) Enantioselective Aza-Henry reaction with acyclic guanidine-thiourea bifunctional organocatalyst. Adv Synth Catal 351(3):345–347. doi:10.1002/adsc.200800692
18. Odagi M, Furukori K, Watanabe T, Nagasawa K (2013) Asymmetric α-hydroxylation of tetralone-derived β-ketoesters by using a guanidine-urea bifunctional organocatalyst in the presence of cumene hydroperoxide. Chemistry 19(49):16740–16745. doi:10.1002/chem.201303006
19. Odagi M, Furukori K, Yamamoto Y, Sato M, Iida K, Yamanaka M, Nagasawa K (2015) Origin of stereocontrol in guanidine-bisurea bifunctional organocatalyst that promotes α-hydroxylation of tetralone-derived β-ketoesters: asymmetric synthesis of β- and γ-substituted tetralone derivatives via organocatalytic oxidative kinetic resolution. J Am Chem Soc 137(5):1909–1915. doi:10.1021/ja511149y
20. Watanabe T, Odagi M, Furukori K, Nagasawa K (2014) Asymmetric α-hydroxylation of a lactone with vinylogous pyridone by using a guanidine-urea bifunctional organocatalyst: catalytic enantioselective synthesis of a key intermediate for (20*S*)-camptothecin analogues. Chemistry 20(2):591–597. doi:10.1002/chem.201303633

21. Cui LQ, Dong ZL, Liu K, Zhang C (2011) Design, synthesis, structure, and dehydrogenation reactivity of a water-soluble *o*-iodoxybenzoic acid derivative bearing a trimethylammonium group. Org Lett 13(24):6488–6491. doi:10.1021/ol202777h
22. Yang T-F, Wang K-Y, Li H-W, Tseng Y-C, Lien T-C (2012) Synthesis of substituted α-tetralones and substituted 1-naphthols via regioselective ring expansion of 1-acyl-1-indanol skeleton. Tetrahedron Lett 53(5):585–588. doi:10.1016/j.tetlet.2011.11.103
23. Schmidt TJ, Vossing S, Klaes M, Grimme S (2007) An aryldihydronaphthalene lignan with a novel type of ring system and further new lignans from linum perenne L. Planta Med 73 (15):1574–1580. doi:10.1055/s-2007-993748
24. Ballini R, Bosica G, Fiorini D, Palmieri A, Petrini M (2005) Conjugate additions of nitroalkanes to electron-poor alkenes: recent results. Chem Rev 105(3):933–971. doi:10.1021/cr040602r
25. Christoffers J, Koripelly G, Rosiak A, Rössle M (2007) Recent advances in metal-catalyzed asymmetric conjugate additions. Synthesis 2007(9):1279–1300. doi:10.1055/s-2007-966005
26. Vicario J, Badía D, Carrillo L (2007) Organocatalytic enantioselective Michael and hetero-Michael reactions. Synthesis 2007(14):2065–2092. doi:10.1055/s-2007-983747
27. Horitsugi N, Kojima K, Yasui K, Sohtome Y, Nagasawa K (2014) Asymmetric Michael reaction of nitroolefins with β-dicarbonyl compounds catalysed by 1,3-diamine-tethered guanidine-thiourea bifunctional organocatalysts. Asian J Org Chem 3(4):445–448. doi:10.1002/ajoc.201402002
28. Sohtome Y, Horitsugi N, Takagi R, Nagasawa K (2011) Enantioselective phospha-Michael reaction of diphenyl phosphonate with nitroolefins utilizing conformationally flexible guanidinium/bisthiourea organocatalyst: assembly-state tunability in asymmetric organocatalysis. Adv Synth Catal 353(14–15):2631–2636. doi:10.1002/adsc.201100219
29. Metcalf WW, van der Donk WA (2009) Biosynthesis of phosphonic and phosphinic acid natural products. Annu Rev Biochem 78:65–94. doi:10.1146/annurev.biochem.78.091707.100215
30. Garzan A, Jaganathan A, Salehi Marzijarani N, Yousefi R, Whitehead DC, Jackson JE, Borhan B (2013) Solvent-dependent enantiodivergence in the chlorocyclization of unsaturated carbamates. Chemistry 19(27):9015–9021. doi:10.1002/chem.201300189
31. Bures J, Dingwall P, Armstrong A, Blackmond DG (2014) Rationalization of an unusual solvent-induced inversion of enantiomeric excess in organocatalytic selenylation of aldehydes. Angew Chem Int Ed 53(33):8700–8704. doi:10.1002/anie.201404327
32. Sohtome Y, Tanaka S, Takada K, Yamaguchi T, Nagasawa K (2010) Solvent-dependent enantiodivergent Mannich-type reaction: utilizing a conformationally flexible guanidine/bisthiourea organocatalyst. Angew Chem Int Ed 49(48):9254–9257. doi:10.1002/anie.201005109
33. Sohtome Y, Shin B, Horitsugi N, Takagi R, Noguchi K, Nagasawa K (2010) Entropy-controlled catalytic asymmetric 1,4-type Friedel-Crafts reaction of phenols using conformationally flexible guanidine/bisthiourea organocatalyst. Angew Chem Int Ed 49 (40):7299–7303. doi:10.1002/anie.201003172

# Index

MIX
Papier aus verantwortungsvollen Quellen
Paper from responsible sources
FSC® C105338

If you have any concerns about our products,
you can contact us on
**ProductSafety@springernature.com**

In case Publisher is established outside the EU,
the EU authorized representative is:
**Springer Nature Customer Service Center GmbH**
**Europaplatz 3, 69115 Heidelberg, Germany**

Printed by Libri Plureos GmbH
in Hamburg, Germany